ELECTRON SPIN RESONANCE SPECTROMETERS

MONOGRAPHS ON ELECTRON SPIN RESONANCE

Editor: H. M. Assenheim, Hilger & Watts Ltd, London

INTRODUCTION TO ELECTRON SPIN RESONANCE, H.M. Assenheim

ELECTRON SPIN RESONANCE IN SEMICONDUCTORS, G. Lancaster

ELECTRON SPIN RESONANCE SPECTROMETERS, T.H. Wilmshurst

Electron Spin Resonance Spectrometers

T. H. WILMSHURST

University of Southampton

Springer Science+Business Media, LLC

Library of Congress Catalog Card No. 68-8257

ISBN 978-1-4899-5599-9   ISBN 978-1-4899-5597-5 (eBook)
DOI 10.1007/978-1-4899-5597-5

Originally published by Plenum Press in 1967.

Softcover reprint of the hardcover 1st edition 1967

CONTENTS

EDITOR'S PREFACE

Dr Wilmshurst's book is intended for physicists and engineers who wish to know more about the working principles of electron spin resonance and its associated circuitry.

This volume covers all facets of E.S.R. instrumentation from the simplest to the most complex spectrometer designs. It is thus a complete guide to all forms of E.S.R. spectrometer, and will enable many to design and construct their own systems from first principles. Classical systems are critically discussed and the author's conclusions embody the most up-to-date techniques and circuitry. There is also very good coverage of the theory behind all parts of the system. In particular, Chapter 4 discusses the design of E.S.R. cavities, resonant and otherwise, so that the non-specialist can now design and construct his own.

For the electronic engineer or microwave engineer who is not concerned with E.S.R., Dr Wilmshurst presents between these covers a complete survey of high-gain low-noise circuitry, A.F.C. for microwave systems, methods of superheterodyne detection, and a multitude of other techniques.

This book forms a most useful addition to the Hilger Monographs on E.S.R., and will be invaluable to any E.S.R. spectroscopist concerned with the practical aspects of his subject. There has long been the need for such a treatment.

H. M. ASSENHEIM

AUTHOR'S PREFACE

In writing this book my purpose has been to set out a systematic approach to the design of an E.S.R. spectrometer. As electron spin resonance is a branch of microwave spectroscopy, the design problem historically has been a matter of translating into terms of microwave hardware the ideas of other branches of spectroscopy. The first chapter traces this development, which draws from optical, infra-red, gaseous microwave and N.M.R. spectroscopy, and also from the field of microwave engineering.

In subsequent chapters I have developed a circuit model for the microwave system and derived an optimum arrangement. Chapter 4 discusses in detail the microwave cavity and covers the problem of wet samples. The remaining chapters are devoted to superheterodyne spectrometers, A.F.C. systems, the magnet and power supplies, and finally, the electronics.

I have assumed that the reader has a graduate knowledge of electronics and microwaves; if he has not, I suggest that he consults some of the books listed in the bibliography.

I am particularly grateful to Dr C.P. Poole, Jr, for allowing me a preview of the breakdown of his book *Experimental Techniques in Electron Spin Resonance*. This has enabled me to avoid duplication, and ideally the present text should be read in conjunction with the other. Dr Poole has provided an excellent and extremely comprehensive review of the various spectrometer designs that have been used. Hence I have limited the scope of my own book to a more detailed discussion of design principles.

Many of the designs described are used in commercial E.S.R. spectrometers. However, as it would have been impractical to list every instance where this is so, I have deliberately avoided mentioning them. I am grateful for the help of a number of manufacturers, who have suggested many of the ideas which I have developed in this book. In particular, I should like to thank the following people: Mr P. Butcher of the Decca Radar Company, who allowed me to spend six weeks with his group at Hersham; Mr H.M. Assenheim of Hilger & Watts Ltd, who has given me full access to the manuals of the Microspin spectrometer and with whom I have had many valuable discussions; and Dr A. Horsfield of Varian Associates, who has allowed me to see some of the instrumentation produced by his company. This is not a comprehensive list of manufacturers, but simply includes those with whom I have had personal discussions.

Thanks are also due to Professor D.J.E. Ingram for introducing me to the subject, and to Mr L.G. Stoodley of R.M.C.S., for innumerable valuable discussions over the past ten years. I am particulary grateful to Mr Assenheim for his careful editorship, stimulation and encouragement, and also to Mr David Tomlinson for seeing the book through the press.

T.H. WILMSHURST
DEPARTMENT OF ELECTRONICS
UNIVERSITY OF SOUTHAMPTON
June, 1967

Chapter 1

Fundamental Requirements

This chapter introduces the subject of E.S.R. spectrometers and indicates their fundamental requirements. First, a brief preliminary discussion of the methods of optical spectroscopy and microwave absorption spectroscopy of gases is given, and then it is shown how the techniques of E.S.R. spectroscopy have developed from them. Enough will be said about the theory of the process of E.S.R. absorption to provide a foundation for the quantitative design of the spectrometer and to give some idea of what kind of spectra to expect. The treatment is necessarily brief and is not intended to be even an introduction to the general topic of E.S.R. spectroscopy. When such an introduction is required it may be found in the first volume[1] in the present series or in other texts listed in the bibliography.

§1.1

OPTICAL SPECTROSCOPY

Optical spectroscopy is the oldest branch of spectroscopy, originating from the simple observation that materials absorb light of some frequencies and transmit light of others. To observe optical absorption spectra, all that has to be done is to pass a pencil of light through the sample and then through a prism. The prism refracts unequally the various frequency components constituting white light, so that when the light falls on a screen different frequency components appear at different points. Frequencies absorbed by the sample produce dark bands or lines on the screen. The frequency resolution of such a system depends on the width of the pencil of light, and so a better arrangement for obtaining good resolution is that of Fig. 1.1. Here, light from the source passes through a narrow slit to a lens which collimates it into a

1

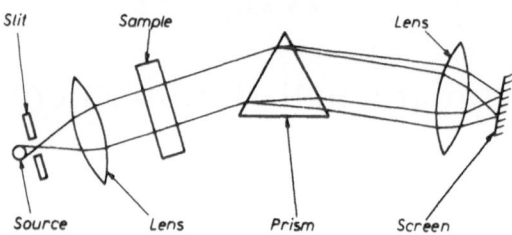

Fig. 1.1. Basic optical spectrometer.

fairly wide, parallel beam. The beam passes through the sample cell and a prism, and finally a second lens focuses it to produce an image of the slit on the screen. With this system, as good a resolution as is required can be obtained by narrowing the slit consistent with the intensity of the source.

The origin of such spectra is intimately connected with the atomic properties of the sample, and to give an adequate description of the spectra quantum mechanics have to be used. Here, the significant difference between classical and quantum mechanics is that with the former a system can adopt any value of energy, whilst with the latter only discrete values or eigenvalues of energy are allowed. Furthermore, the energy of an electromagnetic wave is quantized into small packets or 'photons', and the energy E_p of a photon is strictly related to the radiation frequency by the expression

$$E_p = hf \qquad (1.1)$$

where h is Planck's constant $= 6{\cdot}625 \times 10^{-34}$ joule sec and f the frequency of emitted energy. Thus one way for the atomic system to transfer from one of the allowed energy states to another is to emit or absorb the balance of energy in the form of a photon of the appropriate frequency.

Consequently, it is only possible to exchange energy between a sample and an electromagnetic wave if the frequency of the wave corresponds, according to equation (1.1), to the separation between two of the allowed energy states of the sample. For optical spectra, the appropriate allowed states correspond to the allowed orbits of the electron in an atom. Therefore, when radiation of a frequency corresponding to the separation in energy between one of the inner filled orbits and one of the outer empty ones is applied to the sample, the electrons are stimulated to jump to the outer orbit. As this motion is opposed to the attractive force between the nucleus and the electron, the energy of the outer orbit is greater than that of the inner and so energy is absorbed from the applied field. Because of the relationship between observed spectra and the energy-level scheme of an atom, information about atomic structure can be obtained by studying spectra.

§1.2

MICROWAVE SPECTROSCOPY

The principle of studying the energy-level schemes of physical systems by observing their spectra can be extended to any frequency range and any system of levels. To do this the frequency range of the spectrometer must correspond, according to equation (1.1), to the degree of separation between the energy levels within the system. When spectroscopy was first extended into the microwave region, it was to study the absorption of microwave radiation by gases,[2,3] particularly atmospheric absorption. Fig. 1.2 shows the form of the spectrometer used. A klystron

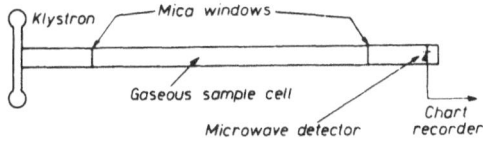

Fig. 1.2. Basic spectrometer for microwave absorption of gases.

microwave oscillator generates electromagnetic radiation, which propagates along a fairly lengthy section of waveguide containing the gas to be examined. The emerging radiation is converted to a d.c. current by a conventional microwave detector. The most important difference between this technique and that of the optical spectrometer is that the klystron, unlike the black-body lamp, is almost a monochromatic source.

Thus the method of using a prism, or its microwave equivalent, to observe a complete set of spectra simultaneously is not appropriate. Instead, the spectra are usually traced out sequentially by varying the frequency of the klystron slowly and by coupling the output from the microwave detector to a chart recorder. The gain in resolution obtained with a monochromatic source is generally agreed to be a more-than-adequate compensation for the need for sequential recording.

By reducing the pressure in the absorption cell, the more-or-less continuous bands of absorption exhibited by certain gases resolve into discrete absorption lines, which can be attributed to transitions between the rotational and inversion levels of the gaseous molecules. By studying the microwave absorption spectra, information can be obtained about these transitions in the same way as information about the orbital states of an electron in an atom can be obtained in optical spectroscopy.

§1.3

ELECTRON SPIN RESONANCE

E.S.R. spectroscopy is also commonly carried out at microwave frequencies. The phenomenon of E.S.R. is most simply explained by considering first the behaviour of a free electron. According to quantum theory the electron is said to be spinning about its axis at one allowed rate which never varies and is the same for all electrons. Consequently, the negative charge that the electron carries is also spinning and constitutes a circulating electric current. The circulating current sets up a magnetic moment which, if the electron is subjected to a steady magnetic field, causes the electron to experience a torque tending to align the magnetic moment with the field. The energy of the system depends on the angle between the magnetic moment and the applied field. Quantum theory stipulates that only two values of energy are permitted, which means that the electron spin can only assume two angles relative to the applied field.

If electromagnetic radiation is applied at a frequency that corresponds, according to equation (1.1), to the separation between the permitted energies, energy is absorbed from the electromagnetic field. This is the phenomenon of E.S.R. The condition for resonance is most readily obtained by assuming that the magnetic moment of the electron is β, the Bohr magneton, and that the moment may be aligned either parallel or antiparallel with the applied magnetic field. It is then readily shown that the energy difference ΔE between the two conditions is given by the equation

$$\Delta E = 2\beta H_0 \qquad (1.2)$$

where H_0 is the value of the applied magnetic field. Thus, from equation (1.1), the frequency f_E at which E.S.R. occurs is given by

$$f_E = 2\beta H_0/h \qquad (1.3)$$

Strictly speaking, the above argument is a little over-simplified. For states of the electron corresponding to the allowed energy values, the electron spin is not exactly parallel or antiparallel with the applied field. Moreover, the magnetic moment of the electron is greater than β. In fact, in both allowed energy states there is an inclination between the magnetic moment and the

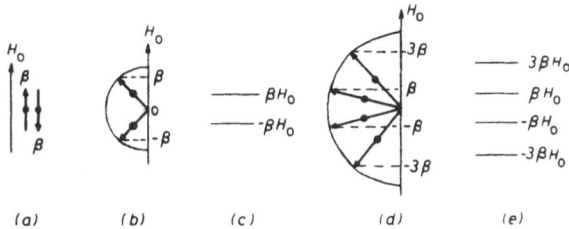

Fig. 1.3. (a) Simplified model of a single electron spin in a steady magnetic field. (b) More exact model for a single electron. (c) Energy-level scheme for a single electron. (d) Model for a three-electron spin system. (e) Energy-level scheme for a three-electron spin system.

applied field, and it is the component of magnetic moment in line with the applied field that is equal to β. The difference between the two views is summarized in Fig. 1.3 where (a) represents the first view and (b) the second. The two models lead to exactly the same energy values, which are indicated in Fig. 1.3(c).

Bound electrons

If we consider electrons bound into an atom, conditions become somewhat different. Often the phenomenon of E.S.R. is not

observed at all, because electrons tend to pair off in the same
way as a collection of bar magnets would if placed in a box and
shaken up. Where pairing is complete, virtually no net spin
magnetism is exhibited and the material is said to be diamagnetic. When pairing is incomplete, the substance is termed 'paramagnetic' and E.S.R. is generally observed. [Another name for
E.S.R. is 'electron paramagnetic resonance' (E.P.R.).] When an
atom has an odd number of electrons, complete pairing is clearly
not possible, but, in such cases, there seems to be a tendency
for the atoms to combine to form molecules in such a way that
the unpaired electrons on the individual atoms can pair together
within the molecule. Therefore, to have an odd number of electrons within the atom is not a sufficient, or even a necessary,
condition to exhibit E.S.R. One important class of materials to
exhibit paramagnetism, and therefore E.S.R., is the transition
elements. In these materials, not only the outer electron shell
but also one or more of the inner shells are only partly filled.
Screening from the outer shells tends to prevent pairing of electrons in the inner unfilled shells with those of adjoining atoms.
More surprisingly, the tendency to pair within the inner unfilled
shell is reversed. In other words, if two electrons are in the
inner shell, they become locked in alignment instead of pairing
off, and for more electrons the tendency, subject to certain re-
strictions, is the same. Thus a group of, say, n electrons behaves as one single electron. Quantum theory states, for the
case of n electrons, that $(n + 1)$ orientations are permissible in
the manner indicated in Fig. 1.3(d) for the case where $n = 3$. In
general, the components of magnetic moment parallel with the
applied field range through the series $n\beta, (n-2)\beta, ..., -n\beta$, and
the corresponding energy levels, illustrated in Fig. 1.3(e), are
split to give a separation of $2\beta H_0$ between adjacent levels.
Selection rules then only allow transitions between adjacent
levels, and the associated frequencies are all the same and
equal to the value for the single-electron resonance.

Few systems behave as simply as this, and when complicat-
ing effects, such as the magnetism arising from the orbital motion
of the electron and the strong electric fields existing in a crystal
lattice, are taken into account, results can be rather different.
Broadly, one finds in such cases that all of the above statements

are violated to a greater or lesser degree. For the single-bound, uncompensated electron, the condition for resonance given by equation (1.3) ceases to be exactly true and the departure is usually expressed by replacing the figure 2 in equations (1.2) and (1.3) by the symbol g, which is then called the spectroscopic splitting factor, i.e.

$$\Delta E = g\beta H_0 \tag{1.4}$$

and
$$f_E = g\beta H_0/h \tag{1.5}$$

For spin systems comprising more than one electron, spacings between the adjacent levels no longer remain equal, so that transitions between different pairs of adjacent levels produce spectral lines at different frequencies. Transitions between levels that are not adjacent are no longer strictly forbidden and so resonances may be observed at frequencies in the region of twice, or several times, the usual values.

Nuclear splittings

A further frequent complication is that nuclei individually associated with the electron spin system often have a magnetic moment. The simplest possible case is that of the nucleus of a single hydrogen atom. The magnetic moment of this nucleus has two allowed orientations, which may again be crudely considered as being parallel and antiparallel with the applied field. The magnetic field associated with the nuclear moment then either adds to or subtracts from the applied field experienced by the electron spin system associated with it. In the bulk sample some electrons will therefore be subject to an increased field and some to a reduced field. Consequently, the original electron resonance line is split into two components spaced equally about the original value for resonance. When nuclei possessing a higher magnetic moment than that of hydrogen are coupled to the electron, multiple orientations of the nucleus are allowed and the electron resonance is split into many lines. Furthermore, when more than one nucleus is present, each line in the multiple spectra due to the splitting of the first nucleus is split again by the second, and so on. Sometimes E.S.R. spectra can be extremely complex.

Choice of operating frequency

It will be clear from equation (1.3) that the **frequency** at which E.S.R. is observed can, in principle, be arranged to have any value by simply adjusting the applied field. To appreciate why frequencies within the microwave range are generally used, it is necessary to examine more closely an assumption tacitly made in §1.1. There it was assumed that transitions were all from inner filled electron shells to outer empty ones and therefore produced absorption of radiation. In general, for any pair of energy levels, E_1 and E_2, within a system, where E_2 is the upper level and E_1 the lower, the relative populations of the levels N_1 and N_2 are given by the Boltzmann distribution

$$\frac{N_2}{N_1} = \exp\left(-\frac{\Delta E}{kT}\right) \tag{1.6}$$

where ΔE is the difference, $E_2 - E_1$, between the upper and lower levels, k Boltzmann's constant $= 1{\cdot}38 \times 10^{-23}$ joules/°C, and T the absolute temperature.

This is simply to say that if the energy separation ΔE between the levels is large compared with the mean thermal energy kT of any element in the system, then the chances of the upper level being populated are remote. On the other hand, if the energy separation is small compared with kT, both energy levels will be almost equally populated. For optical spectroscopy at room temperature, ΔE is in the region of 4×10^{-19} joules whilst $kT \simeq 4 \times 10^{-21}$ joules, so that the upper levels are largely unoccupied. For microwave spectroscopy, however, frequencies are in the region of 10 kMc/s, so $\Delta E \simeq 6 \times 10^{-24}$ joules, and by equation (1.6) the two spin levels will be nearly equally populated.

The interaction between the wave and the sample then becomes more complex. Electrons in the lower energy level are stimulated upwards and absorb photons of energy in the ordinary way, whilst electrons in the upper energy level are stimulated downwards and emit energy. It is the excess of absorption, due to the slightly greater population of the lower level, that is observed and this excess should therefore be made as large as possible. Again from equation (1.6), the higher the value of ΔE, the greater the difference between N_1 and N_2. As $\Delta E \propto H_0$ [see equation (1.2)], the applied magnetic field should be as high as possible. In fact, the order of field (3 kilogauss) that can conveniently be provided by a good laboratory magnet gives a frequency in the microwave region. Moreover, for reasons to be discussed in Chapter 4, there is little point in attempting to increase this field value. Hence E.S.R. spectroscopy is most commonly, but not always, carried out at microwave frequencies.

§1.4

THE E.S.R. CAVITY

Experimentally one of the main differences between E.S.R. spectroscopy and the microwave absorption spectroscopy of gases is that E.S.R. requires a fairly intense applied magnetic field whilst gaseous spectroscopy needs no field. The requirement of a field poses certain sensitivity problems. For optical spectroscopy it would seem a general rule that the longer the path of the beam of light through the sample the more sensitive the system. This argument is not quite true, and when translated into the terms of a gaseous microwave spectrometer gives the result (§3.5) that the length of waveguide required for optimum sensitivity is that which will give a power loss of exp (–2) to a wave passing through it. For a standard copper waveguide with an attenuation of 5·5 dB/100 ft, the length required would be over 100 ft, which is clearly impossible to accomodate between the poles of a normal magnet. The most common technique then is to

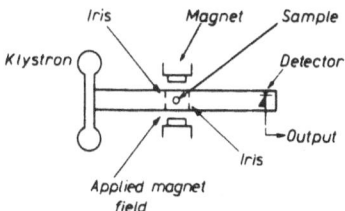

Fig. 1.4. E.S.R. spectrometer with resonant cavity.

place the sample in a resonant microwave cavity. Fig. 1.4 shows a simple arrangement where the resonant cavity is formed by placing two metal irises across the waveguide. The irises must be an integral number of half-guide wavelengths apart and must have small holes or slots in them to allow the microwave energy to be coupled in and out. The resonant cavity is a solution

suggested by previous experience in various fields, notably
nuclear magnetic resonance, infra-red spectroscopy and general
microwave engineering. One way of understanding the transition
from the travelling wave absorption cell to the resonant cavity is
to say that the pathlength is restored to the required value by
repeated bouncing of the wave between the two irises. This is
true, but does not sufficiently emphasize the need for the cavity
to be resonant. In order to obtain many reflections it is necessary that the transmission through the coupling hole should be
small. On the other hand, for energy from the microwave generator
to flow into the cavity the iris should be removed altogether. If
the cavity is non-resonant these considerations cannot be re-
conciled, but if the irises are an integral number of half-wavelengths apart, it is possible to use very small coupling holes and,
at the same time, to couple energy into and out of the cavity
efficiently. Suppose for the moment that the coupling hole in the
first iris is such that it allows a very small transmission in the
ordinary way. Most of the initial incident wave from the microwave generator is then reflected back to the generator. The small
transmitted wave proceeds to the second iris, is reflected almost without any transmission loss, and returns to be reflected
again at the first iris. Because the irises are an integral number
of half-wavelengths apart, the wave transmitted through the iris
from the generator at the moment of second reflection is exactly
in phase with the doubly reflected wave. The two waves proceed
forwards mutually reinforcing each other. Subsequently, the wave
bouncing between the two irises is built up from the individual
small components of the wave that the generator transmits through
the coupling hole.

As the intensity of the bouncing or 'standing' wave increases,
so waves of increasing intensity are transmitted out to the detector and back to the generator. A proper analysis shows that
the wave propagated back to the generator is exactly in antiphase with the wave reflected from the first iris. Consequently
the effect is observed as a reduction in the reflected wave. In
fact, by suitably adjusting the size of the coupling hole, it is
possible to make the two waves equal and opposite so that no
net component is reflected back to the generator. Analysis in
subsequent chapters will show that this is not exactly the most

sensitive condition in which to operate the cavity for E.S.R. spectroscopy, but it is at least clear that, if no wave is being reflected back to the generator, all of the available energy is being coupled through the first iris and into the cavity. Briefly, the effect of tuning the cavity to resonance is that when the frequency of the generator is adjusted so that the separation between the irises corresponds to an integral number of half-wavelengths, the intensity of the reflected wave is reduced and may become zero; then a strong standing wave is built up in the cavity and a travelling wave is propagated in the direction of the detector. When the cavity is off-resonance, most of the wave from the generator is reflected directly back from the first iris.

Many other forms of cavity resonator exist, notably cylindrical, coaxial and strip-line resonators which will be discussed in Chapter 4.

§1.5

THE OPERATION AND SETTING UP OF THE KLYSTRON

As the klystron oscillator is of fundamental importance to the operation of an E.S.R. spectrometer, it is appropriate to give a very crude, over-simplified view of its operation. A more exact treatment is given by Reich *et al.*[4] and a thorough discussion by Hamilton.[5] A sketch of a reflex klystron is shown in Fig. 1.5.

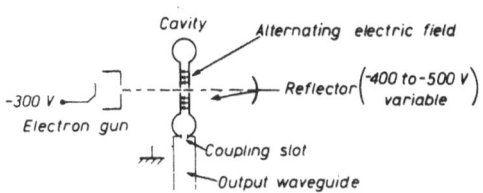

Fig. 1.5. Elementary diagram of klystron oscillator.

Here, an electron gun emits a beam of electrons which passes through a hole in the microwave cavity. The mode of operation of the cavity is such that the alternating electric field is across the centre of the gap as shown. The electrons are therefore alternately accelerated and decelerated according to the phase of the cavity oscillation. No immediate change in the density of the electron beam occurs, but the periodic modification in the velocity of the emerging electrons causes the electrons to bunch together at a point beyond the plane of the cavity. The bunching

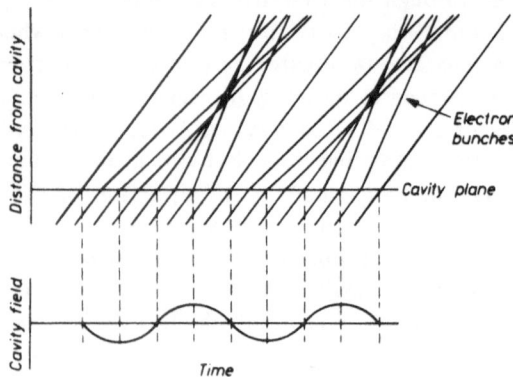

Fig. 1.6. Applegate diagram showing formation of first bunching plane in the electron beam of a klystron.

is illustrated very clearly by the well-known Applegate diagram of Fig. 1.6. Here each line represents the progress of an individual electron in a time-position graph. Before reaching the cavity the electrons are arriving at a constant rate, all with the same velocity (which is primarily determined by the positive potential applied between the cavity and cathode). After passing through the cavity, the velocity, and therefore the slope of the electron trajectories on the diagram, is periodically altered to give the bunching effect shown. The bunches then persist for a considerable distance before dispersing.

As some electrons are accelerated and some decelerated, the bunching process, generally, extracts no energy from the cavity. If however a negative potential is applied to the reflector relative to the cavity, a strong retarding field will be applied to the electron beam. Moreover, if the reflector potential is below that of the cathode, the electrons will be brought to a halt before reaching the reflector and will be transmitted back to the cavity. If the reflector potential can be adjusted so that the phase relationship between the bunches and the cavity oscillations is such that the bunches experience a retarding field on passing through the cavity, then the bunches will be slowed, energy will be extracted from the beam by the cavity, and the oscillations will be sustained. Energy is coupled from the cavity to the waveguide by a slot or hole in the cavity, or by a coupling loop. The above mode of operation demands a reflector voltage of a particular value, or at least one of several values, which brings the bunches through the cavity at the appropriate stage of the cavity oscillation. If the klystron output is plotted against the reflector volt-

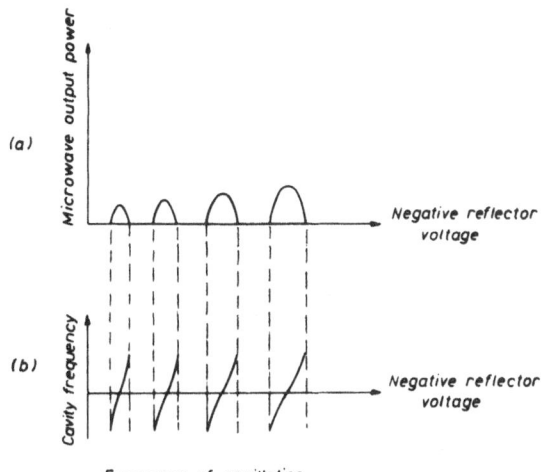

Fig. 1.7. Variation of output power and frequency of oscillation with reflector voltage for a reflex klystron oscillator.

age, a graph of the form shown in Fig. 1.7(*a*) results. This graph shows that oscillations are sustained over a fairly wide range of reflector voltages round about the required values. If the klystron frequency is plotted in the same way [see Fig. 1.7(*b*)], the frequency varies from somewhat below the resonant frequency of the cavity to somewhat above it as the reflector voltage becomes increasingly negative. The range of frequencies over each span of reflector voltage for which oscillation occurs is the same and the reason for 'pulling' is that, if the reflector voltage is too strongly negative, the bunch arrives a little too soon for the cavity and so tends to increase the frequency of oscillation. Conversely, if the reflector voltage is not negative enough, the frequency is lowered. One method of illustrating these effects is to couple the klystron to a crystal detector via an absorption

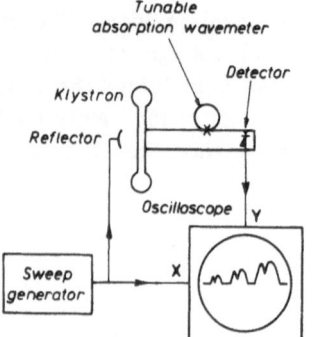

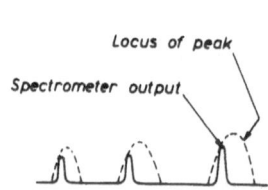

Fig. 1.8. Use of absorption wavemeter to measure variation of klystron frequency.

Fig. 1.9. Detected output from E.S.R. spectrometer as reflector voltage is swept.

wavemeter, as shown in Fig. 1.8. In addition to the normal d.c. voltage, a sweep voltage is applied to the reflector and also to the X-plates of an oscilloscope. The output from the detector is coupled to the Y-plates of the oscilloscope to give a display corresponding to Fig. 1.7(a). When the absorption wavemeter is tuned to a frequency covered by the klystron, a dip is observed on each klystron 'mode', as shown in Fig. 1.8. If the wavemeter is tuned, this dip is observed to traverse each klystron mode simultaneously. It is also possible to observe the same effect by adjusting the dimensions of the klystron cavity.

For operating the E.S.R. spectrometer it is desirable that maximum output should be obtained from the klystron, and so the reflector voltage should be adjusted to the top of the largest mode. The largest mode is generally that corresponding to the most negative reflector voltage, for then the bunches reach the cavity most rapidly and therefore have the least time to disperse.

If the output from the detector of the spectrometer of Fig. 1.4 is displayed in the same way, the dip of the absorption wavemeter will be replaced by a transmission peak, as shown in Fig. 1.9. If the resonant frequency of either the spectrometer cavity or the klystron cavity is adjusted, the transmission peaks observed on the oscilloscope will trace out the form of the klystron mode. It is then a simple matter to make adjustments so that the spectrometer cavity resonance coincides with the peak of the klystron mode. Once this has been done, the sweep can be removed from the reflector whilst, at the same time, adjusting the d.c. reflector voltage to keep the required peak in the centre of the oscilloscope screen. When the sweep is eventually removed, a steady vertical deflection is observed, which will decrease if the klystron reflector voltage is adjusted in either direction. As it is possible to set up the spectrometer by adjusting the klystron cavity, it is quite common for the spectrometer cavity not to be tunable.

A convenient method for supplying the sweep to the reflector voltage is shown in Fig. 1.10. Here sweep of an amplitude controlled by R_v is applied via the network of C and R When the sweep is removed the CR network then provides additional

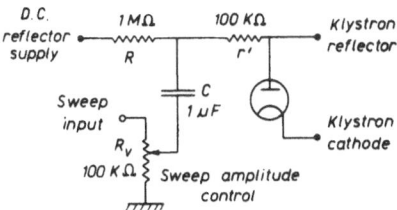

Fig. 1.10. Method for applying a sweep voltage to the reflector of the klystron.

smoothing of the reflector supply. The diode and resistor r' ensure that the sweep voltage can never drive the reflector positive with respect to the cavity, otherwise the reflector would begin to pass a current, overheat and quickly be destroyed.

Frequency and power monitoring arrangements

For reference purposes it is often important to know what power is being delivered by the klystron to the spectrometer, or at least to be able to keep the power constant or to vary it by a known amount. Consequently, it is usual to adopt the arrangement shown in Fig. 1.11. Here a directional coupler feeds a constant fraction (usually –10 dB) of the output from the klystron to a monitoring detector. A bolometer is initially inserted at point X, the power adjusted by the padding attenuator to some conveni-

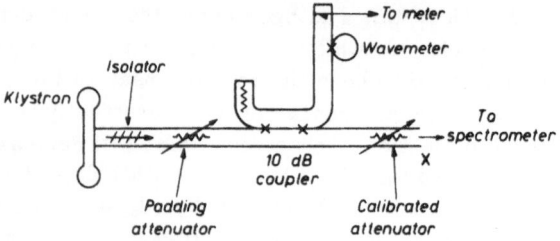

Fig.1.11. Normal method of monitoring power and operating frequency of a spectrometer klystron.

ent value, and the output from the monitoring detector noted. Subsequent variations in the output from the klystron are then corrected by the padding attenuator, which is adjusted to keep the output from the monitoring detector constant. Any required reduction from the maintained output power is then provided by the calibrated attenuator.

An absorption wavemeter is usually inserted in the side arm, so the klystron frequency can be measured by observing at what setting of the wavemeter the monitor output dips.

It is also normal to place a ferrite isolator [6] immediately after the klystron. The isolator is a device which allows waves to propogate in one direction but not in the other. In the reverse direction the waves are absorbed. By using the isolator any reflections from the spectrometer are prevented from reaching the klystron. Generally such reflections tend to pull the klystron frequency, particularly when adjustments are being made to the cavity etc.

References

1. Assenheim, H.M., *Introduction to Electron Spin Resonance* (Hilger & Watts, 1966).
2. Townes, C.H., and Schawlow, A.L., *Microwave Spectroscopy* (McGraw Hill, 1955).
3. Gordy, W., Smith, W.V., and Trambarulo, R.F., *Microwave Spectroscopy* (Wiley, 1953).
4. Reich, H.J., Skalnik, J.G., Ordung, P.F., and Krauss, H.L., *Microwave Principles* (Van Nostrand, 1957).
5. Hamilton, J.J., *Reflex Klystrons* (Chapman & Hall, 1958).
6. Waldron, R.A., *Ferrites* (Van Nostrand, 1961).

Chapter 2

Development of a Basic Spectrometer

In the previous chapter we arrived at a design for an E.S.R. spectrometer (Fig. 1.4) by considering how the instrumentation of some of the previously developed branches of spectroscopy can be modified to suit the particular frequency range and application of E.S.R. Here we shall discuss some of the methods whereby this and similar arrangements can be used to trace out spectra. Then from the Bloch model of magnetic resonance we shall develop a circuit model for the microwave system, and from this shall derive an improved arrangement. Finally, some modifications to the improved system will be suggested in the light of considerations of detector sensitivity and the need to discriminate between the absortion and dispersion components of the E.S.R. signal.

§2.1

BASIC RECORDING METHODS

Since the condition for magnetic resonance given by equation (1.5) involves the field H_0 applied by the magnet and the frequency f of the microwave generator, it is, in theory, possible to trace a spectrum by varying either of these parameters. Thus, if the output of a simple spectrometer is connected to a chart recorder which has the paper moving at a constant speed, and if, in principle, either the field or the frequency is varied in a linear fashion, then a spectrum can be traced on the chart. When a resonant cavity is used, however, it is only possible to vary the field, because the cavity does not remain at resonance when the frequency is varied. The effect of this is twofold. Firstly, the sensitivity of the spectrometer varies considerably over the extent of the spectrum and, secondly, the response curve of the

17

cavity is traced on top of the spectrum. Furthermore, the latter effect represents a 100 per cent variation in the output of the microwave detector, and, since the system can otherwise detect fractional variations of many orders below unity, most potentially observable E.S.R. signals are lost.

The next limitation to sensitivity is that the gain of any d.c. amplifier included prior to the chart recorder cannot be any greater than that which drives the recorder pen off-scale. This difficulty is simply overcome by connecting a d.c. balancing source in series with the output from the detector, and adjusting the balancing potential so that it is equal and opposite to the normal output. The steady direct current to the recorder is then zero, so that, in principle, any degree of d.c. amplification can be used prior to the recorder without overloading the amplifier or driving the pen off-scale.

In practice, the degree of amplification so obtained is limited by the effects of drift. Several sources of drift exist in the system suggested and the first is the d.c. amplifier itself. In the d.c. amplifier small changes in the current drawn by the earlier stages are amplified by the later stages and cause the pen recorder to wander across the chart and obscure the E.S.R. signal. The cause of these changes is generally thermal, due initially to warming up after switching on and subsequently to changes in ambient temperature. Further sources of drift appear as changes in the output from the microwave detector. These changes partly originate within the detector itself and are partly due to changes in the amplitude of the signal generated by the klystron. Generally the most serious contribution to drift is thermal drift in the klystron frequency. This frequency drift is translated to drift in the level of the signal reaching the microwave detector because of the frequency response curve of the spectrometer cavity.

§2.2

CRYSTAL-VIDEO RECORDING

The crystal-video spectrometer of Fig.2.1 overcomes the problem of drift. Here, instead of varying the field slowly, a periodic field sweep is applied to the sample by winding a pair of sweep coils onto the pole pieces of the magnet and driving them from

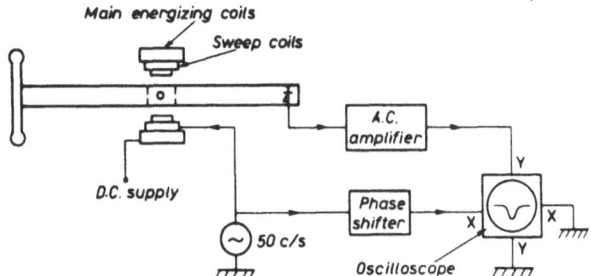

Fig. 2.1. Crystal-video spectrometer.

the 50 c/s mains supply. The absorption spectrum appears at the output of the microwave detector in the form of a periodic 50 c/s waveform, which can be amplified by an a.c. amplifier that does not respond to d.c. current and, therefore, to drift in the d.c. level. The output from the amplifier is displayed on an oscilloscope, which uses the supply that provides the field sweep to give the X-deflection. Since the field-sweep is sinusoidal, it is necessary to use the field-sweep directly, and not as a means of synchronizing the normal linear time base of the oscilloscope. If the latter approach is adopted, distortion of the displayed waveform will result. Even when the sinusoidal X-sweep is used, it is necessary to include a phase shifter in the system, otherwise the oscilloscope responds to the voltage applied to the sweep coils, whilst the sweep itself is proportional to the current through the coils, which is not in phase with the voltage. Phase correction would be unnecessary if the resistance of the coils were much larger than the reactance; then the voltage and current would be in phase. In practice, however, the reactance is much larger than the resistance, so a phase correction approaching $90°$ is required.

The main factor that limits the sensitivity of the crystal-video spectrometer is the very fundamental one of 'noise'. This electrical 'noise', taking the form of small fluctuating voltages or currents, occurs in any electronic circuit, and in the crystal-video spectrometer appears as a random, 'grass-like' deflection of the trace. A typical trace is that of Fig. 2.2, where the noise can be seen superimposed on a normal single-line absorption spectrum.

The primary sources of noise are thermal or 'Johnson' noise, shot noise, partition noise, and flicker noise. Thermal noise is another name for black-body radiation and originates from the thermal motion of the charge carriers in a conductor or semiconductor. Shot noise occurs when carriers have to traverse a junction, such as the P–N junction in a diode or transistor, or the anode-cathode space of a valve. As the carriers are discrete, the current is not continous and, in reality, comprises a series

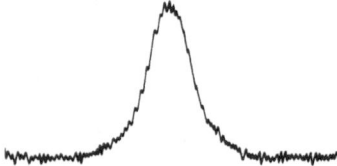

Fig. 2.2. Typical output display from a crystal-video spectrometer, showing noise.

of tiny pulses. These pulses generally average out to give a steady d.c. component, but a small fluctuation remains which constitutes the effect. Partition noise is of less importance now that solid-state circuitry is being used more and more. It is due to a further degree of randomness introduced when an electron has to choose between, say, going to the anode or to the screen grid of a pentode. These three types of noise differ in one respect from flicker noise, whose origin is still poorly understood. They are all 'white' noise, that is, noise whose spectral intensity is independant of frequency. Flicker noise, on the other hand, has a spectral intensity inversely proportional to frequency, and is sometimes known as $1/f$ noise. It is normal to denote the amplitude of a noise voltage or current by its root mean square (r.m.s.) value, e.g. $(v_n^2)^{1/2}$ or $(i_n^2)^{1/2}$, or, more concisely, $\tilde{v}_n$ or $\tilde{i}_n$. For a constant spectral intensity, the noise power admitted through a bandpass filter is proportional to the bandwidth, so that, for white noise v_n^2 is proportional to bandwidth and independent of frequency. For flicker noise, v_n^2 is inversely proportional to frequency. Because white and flicker noise are usually present in an electronic device, and because one is independent of frequency and the other is not, there is always a transition frequency above which the white noise predominates and below which the flicker noise predominates. For frequencies above the transition frequency the spectral intensity of noise is constant, and for frequencies below the transition point the spectral intensity rises with $1/f$. For valve and transistor amplifiers the transition frequency is in the region of 10 kc/s, whilst for conventional silicon microwave detectors the value is approximately 10 Mc/s. Consequently, for a crystal-video spectrometer operating at 50 c/s, the noise from both crystal and amplifier is, without doubt, predominately flicker noise.

It would therefore seem advantageous to operate the system with as high a sweep frequency as possible, so that the response of the a.c. amplifier can be raised to a point in the spectrum where flicker noise is reduced. Unfortunately, the advantages of this method are cancelled by the need for an increased bandwidth. The waveform at the output of the microwave detector can be represented by a Fourier series with a fundamental frequency equal to that of the field modulation f_m, and with significant harmonic components up to about $100f_m$. Thus the bandwidth of the a.c. amplifier must extend at least from f_m to $100f_m$. However, if for normal circuitry the phases of the fundamental component and of the lower harmonics are not to be shifted relative to the majority of the harmonics, the low-frequency response of the amplifier must generally be maintained down to about $f_m/100$. Thus there is a need for a total band-pass ranging from $f_m/100$ to $100f_m$, or more generally f_m/a_D to a_Df_m, where a_D is a factor depending on the degree of distortion that can be tolerated. It should perhaps also be mentioned that for symmetrical spectra — as they are more often than not — the fundamental frequency of the waveform is $2f_m$ and, more correctly, the band-pass of the amplifier should range from $2f_m/a_D$ to $2a_Df_m$. If the spectrum contains n_l lines, the upper limit must be raised by the factor n_l.

These finer points do not influence the main argument that, if f_m is raised by any given factor, the upper and lower limits of the band-pass of the amplifier must similarly be raised, and so must the bandwidth. Thus the gain obtained by reducing flicker noise with frequency is exactly counterbalanced by the required increase in bandwidth. Stating this argument more precisely, one may express the spectral intensity $W(f)$ of the flicker noise, i.e.

the noise power per unit bandwidth, by the relation

$$W(f) = \frac{A_n}{f} \qquad (2.1)$$

where A_n is a constant. The total noise power at the output of the amplifier is then given by

$$P_n = \int_{2f_m/a}^{2n_l a_D} \frac{A_n}{f} \cdot df$$

$$= A_n \ln(n_l a_D{}^2) \qquad (2.2)$$

As this expression is independent of f_m, there is no advantage in altering the modulating frequency in either direction.

§2.3
DOUBLE-MODULATION RECORDING

The technique of double-modulation recording avoids the un-favourable connection between bandwidth and modulating frequency that occurs when using a crystal-video spectrometer. Double-modulation recording (Fig. 2.3) differs from crystal-video detection in that the slow linear field-sweep used in the

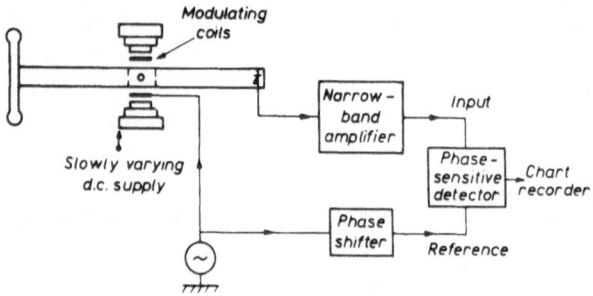

Fig. 2.3. Spectrometer for double-modulation recording.

elementary recording technique is restored, whilst the amplitude of the a.c. sweep used in crystal-video recording is reduced until its extent is small compared with the width of a spectral line. The waveforms obtained in double-modulation recording are given in Fig. 2.4. Fig. 2.4(a) represents the change in output at the detector (much exaggerated) as a simple function of the steady magnetic field at the sample, (*b*) shows the variation of field as a function of time when the linear and sinusoidal field variations are applied, and (*c*) shows how the static curve of

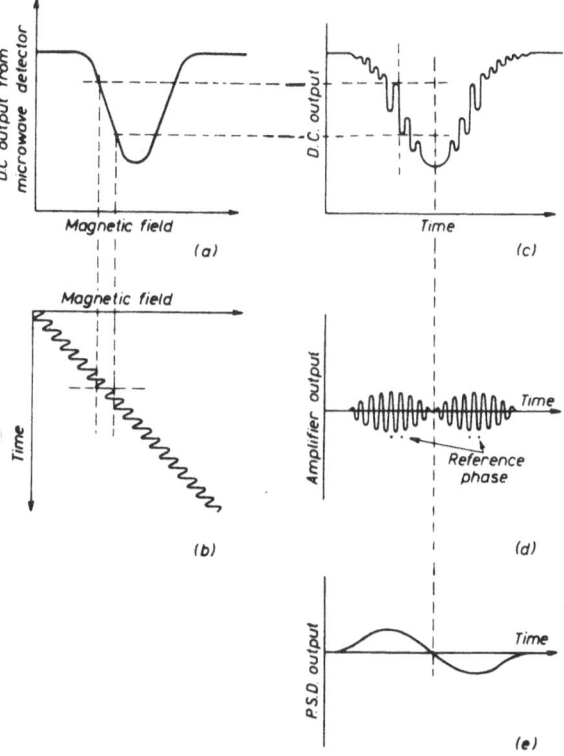

Fig. 2.4. Waveforms illustrating the operation of a double-modulation spectrometer. (*a*) Variation of output from microwave detector with applied magnetic field. (*b*) Variation of magnetic field with time. (*c*) Resulting variation in output from microwave detector with time. (*d*) Output from narrow-band amplifier with time. (*e*) Output from phase-sensitive detector with time.

(*a*) translates these variations into a time-varying signal at the output of the microwave detector. A sinusoidal component proportional to the slope of the absorption line is superimposed upon the output obtained by the elementary recording technique, and the combined waveform is fed to a narrow-band amplifier tuned to accept only the sinusoidal component. The signal at the output of the amplifier varies with time in the manner indicated in Fig. 2.4(d). The change in sign of the derivative of the absorption curve is conveyed in the waveform of (d) as a reversal in phase relative to the field modulation waveform. Consequently, if the waveform at the output of the amplifier is fed to a phase-sensitive detector (P.S.D) that uses the original field-sweep waveform as reference, the d.c. output from the detector will be a faithful reproduction of the derivative of the absorption line, as shown in Fig. 2.4(e), and can be fed to the pen recorder. The reasons for having the phase shifter are the same as those for its inclusion in the crystal-video spectrometer.

A simple form of phase-sensitive detector which serves to demonstrate the principle of the device is shown in Fig. 2.5. Here the reference signal opens and closes the contacts of a relay, which determines whether or not the input signal appears at the output. For the phase relationship between signal and reference [see Fig. 2.6(a)], the output is always positive, but if the phase

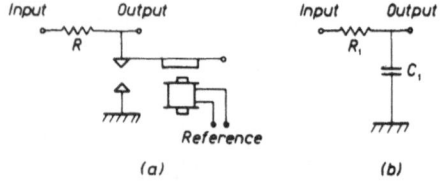

Fig. 2.5. (a) Basic phase-sensitive detector. (b) Smoothing network. ($R_1 \gg R$, $R_1 C_1 \gg$ signal period.)

of the signal is reversed, as in Fig. 2.6(b), the output is always negative. Consequently, if the resulting output waveform is smoothed by a low-pass filter, such as that shown in Fig. 2.9(b), the first case will give a steady positive output and the second a steady negative output. Here $R_1 \gg R$ to prevent the filter from loading the phase-sensitive detector. To obtain adequate smoothing $R_1 C_1$ must also be made much greater than the period of the signal. More generally, it is shown below that, if ϕ_p is the phase between the signal and reference, the d.c. output is proportional to $\cos \phi_p$. For example, if $\phi_p = 90$, the output is zero. This is clearly illustrated in Fig. 2.6(c), where the output waveform is as often negative-going as it is positive. A phase shifter is therefore

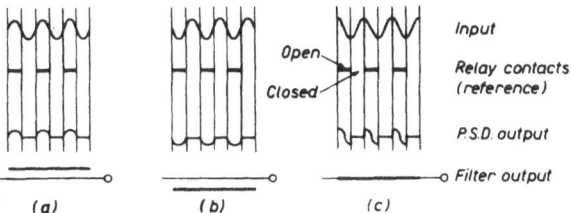

Fig. 2.6. Phase-sensitive detector waveforms. (*a*) Input and reference in phase. (*b*) Input and reference in antiphase. (*c*) Input and reference in quadrature.

required to ensure that the reference is correctly phased relative to the signal.

The bandwidth required for double-modulation recording is approximately equal to the inverse of the time taken to trace out one line of an E.S.R. spectrum. This can be shown by considering the effect of the bandwidth either of the narrow-band amplifier or of the d.c. system following the phase-sensitive detector. The amplitudes of the components of the Fourier series for a regular train of pulses of width ΔT and period T_p remain approximately constant for the lower-frequency components ranging from the fundamental component of frequency T_p^{-1} to components of frequency $(\Delta T)^{-1}$, whilst the amplitudes of the higher-frequency components decay to zero. If T_p is allowed to tend to infinity, whilst ΔT remains constant, so that a single pulse is represented, the Fourier series becomes a Fourier integral representing an infinite number of infinitely small sinusoidal components extending from zero frequency to $(\Delta T)^{-1}$. Thus, to amplify the pulse adequately without distortion, the bandwidth of the amplifier must be somewhat greater than the inverse of the pulse width. To amplify the transient output from the phase-sensitive detector of the spectrometer, corresponding to the derivative of a spectral line, the bandwidth of the d.c. amplifier must be greater than the time taken to trace out a spectral line.

A more precise estimate of the degree of distortion that is likely to result from a restriction in bandwidth is obtained by considering the transient response of the network of Fig. 2.5(*b*) to a step function, i.e. a function which is zero for $t < 0$ and a constant value A_s for $t > 0$. For $t > 0$, the mesh equation for

Fig. 2.5(b) is

$$A_s = iR_1 + v_0 \tag{2.3}$$

where i is the current and v_0 the output voltage across C_1. As the charge q on the capacitor is equal to $C_1 v_0$ and $i = dq/dt$, equation (2.3) may be written

$$A_s = C_1 R_1 \frac{dv_0}{dt} + v_0 \tag{2.4}$$

which has the solution

$$v_0 = A_s \left[1 - \exp \frac{-t}{C_1 R_1} \right] \tag{2.5}$$

if $v_0 = 0$ when $t = 0$. Thus the rise in output voltage is exponential, with a time constant of $C_1 R_1$.

The bandwidth of the network can then be obtained from the complex transfer function, which gives the output voltage V_0 as a function of the frequency of a sinusoidal input voltage V_{in}. Clearly, V_0 / V_{in} is equal to the ratio of the capacitive reactance to the total impedance of the circuit, i.e.

$$\frac{V_0}{V_{in}} = (1 + j\omega C_1 R_1)^{-1} \tag{2.6}$$

Generally, the cut-off frequency of a network is defined as the frequency at which its response falls by a factor of $\sqrt{2}$ from its maximum value. Here it is clear that $\left| V_0 / V_{in} \right| = \left| V_0 / V_{in} \right|_{\omega=0}$ $\times 1/\sqrt{2}$ when $\omega = 1/C_1 R_1$, and so the cut-off frequency $f_c = \omega_c / 2\pi$ is given by

$$f_c = (2\pi C_1 R_1)^{-1} \tag{2.7}$$

The above analysis is included to emphasize the often forgotten factor of 2π in equation (2.6). The bandwidth that will place an exponential rise with a time constant of $C_1 R_1$ on the edge of a step is not $1/C_1 R_1$ but $1/2\pi C_1 R_1$. In other words, the bandwidth actually required is $1/2\pi$ times less than is commonly supposed.

It is also important to determine the bandwidth requirement for the narrow-band amplifier in a double-modulation recording system. The effect of double modulation is to present to the ampli-

fier a sinusoidal signal of a frequency equal to the field-modulating frequency f_m and of amplitude proportional to the time varying derivative of the E.S.R. spectrum. Consequently, the original
sine wave is amplitude-modulated by a transient which has frequency components extending from zero to the inverse of the time
taken to traverse one spectral line. If only one modulating component of frequency f' is considered, the voltage v at the input
of the amplifier is given by

$$v = \hat{v} \cos \omega' t . \cos \omega_m t \qquad (2.8)$$

where $\hat{v}$ is the peak value of v and constant, $\omega' = 2\pi f'$, and
$\omega_m = 2\pi f_m$. Equation (2.8) can then be expanded to

$$v = \frac{\hat{v}}{2} \cos(\omega_m - \omega')t + \frac{\hat{v}}{2} \cos(\omega_m + \omega')t \qquad (2.9)$$

that is, into two sinusoidal sideband components situated on
either side of ω_m and separated from it by a space ω'. When all
components of the transient are considered, it becomes clear
that the signal presented to the input of the narrow-band amplifier is centered at the field-modulating frequency f_m and has a
spread of approximately twice the inverse of the time taken to
trace out one spectral line. Thus, the bandwidth required of the
narrow-band amplifier is twice that of the d.c. amplifier.

The most important point about the bandwidth of a double-
modulation-recording system is that it is not linked to the modulation frequency as the bandwidth of the crystal-video spectrometer is. Hence the modulation frequency can be increased to
reduce the effects of flicker noise, whilst to give a further reduction in noise the bandwidth of the system may be reduced, provided that the rate of recording is suitably lowered. For practical
reasons, which will be discussed later (§3.14), some compromise
has to be reached when deciding the frequency to which the field
modulation can be raised. A value of 100 kc/s is a common
choice.

Comparing the sensitivities of typical crystal-video and double-
modulating systems, an increase in f_m from 50 c/s to 100 kc/s
gives a reduction in flicker noise power of 2×10^3, and, if the
sweep rate is such that one line is traced out in, say, 10 sec,
then a bandwidth of 1 c/s provides a comfortable margin for

avoiding distortion. Compared with the figure of approximately 10 kc/s required for crystal-video recording, a further reduction of 10^4 in noise power is thus obtained.

It should be emphasized that the final improvement of 2×10^7 is a reduction in noise *power* whilst the recorder deflection is proportional to noise and signal *voltage*, so the improvement in signal-to-noise ratio expressed as a recorder deflection is $4\cdot5 \times 10^3$. The above calculation assumes that the spectral intensity at 50 c/s remains constant at all frequencies within the band-pass of the crystal-video spectrometer. This is not true for flicker noise. When equation (2.1) is used, one has for double-modulation recording

$$P_n = \int_{10^5}^{10^5+1} \frac{A_n}{10^5} \cdot df$$

whilst for the crystal-video system, using equation (2.2) and having $n_l = 1$ and $\alpha_D = 100$, one obtains

$$P_n = A_n \ln (100^2)$$

which gives an improvement in voltage signal-to-noise ratio of only approximately 10^3. More generally, the improvement obtained by double-modulation recording is given by the ratio S_i, where

$$S_i = \ln(n_l \alpha_D^2) \cdot \frac{f_m}{B} \tag{2.10}$$

B and f_m being, respectively, the bandwidth and the modulating frequency of the double-modulation system, and assuming the noise intensity to be strictly proportional to $1/f$.

The practical problem of providing at the sample a modulating field with a frequency of 100 kc/s involves some difficulties. If the modulating coils are outside the cavity, the cavity walls must be extremely thin to prevent shielding (see §4.4). Another method, now not so favoured as the one just mentioned, requires a small loop to be placed inside the cavity and around the sample.

In order to proceed with a practical design for the narrow-band amplifier and phase-sensitive detector for a double-modulation-recording spectrometer, it is necessary to discuss the simple mathematical theory of the phase-sensitive detector. Initially we assume that the input signal v_{in} delivered by the narrow-band amplifier is entirely monochromatic and so can be represented by the expression

$$v_{in} = \hat{v}_{in} \sin\omega t$$

where $\hat{v}_{in}$ is the peak value of v_{in} and $\omega = 2\pi f$, f being the frequency of the signal, which, for the moment, is assumed not to equal f_m. We can express the switching action of the phase-sensitive detector by considering the input signal to be multiplied by a function $h(t)$, which is unity when the switch is open and zero when the switch is closed. $h(t)$ expressed as a Fourier series is given by

$$h(t) = \frac{1}{2} + \frac{2}{\pi}\left(\sin\omega_m t + \frac{\sin 3\omega_m t}{3} + \frac{\sin 5\omega_m t}{5} + \dots\right)$$

where $\omega_m = 2\pi f_m$.

The output voltage v_0 therefore becomes

$$v_0 = v_{in} h(t)$$

which may be expanded into a series of single-frequency components to give

$$v_0 = \hat{v}_{in}\left\{\frac{\sin\omega_m t}{2} + \frac{1}{\pi}\left[\cos(\omega_m - \omega)t - \cos(\omega_m + \omega)t\right] + \frac{1}{3\pi}\left[\cos(3\omega_m - \omega)t - \cos(3\omega_m + \omega)t\right]\dots\right\} \qquad (2.11)$$

This result is illustrated in Fig. 2.7(a-c), where (a) shows the signal frequency f, (b) the switching terms and (c) the output components. For the most part, the output components occur at frequencies which are separated by the signal frequency f on either side of the frequency of each switching term, and are subject to the decrease in amplitude, with the order of the switching term from which they stem. The only exception is the zero-frequency switching term, which stems only in the positive direction. For the E.S.R. signal, the input frequency f is equal to the modulating frequency f_m, so that the term $(\hat{v}_{in}/\pi)\cos(\omega_m - \omega)t$ becomes $\hat{v}_{in}/\pi$ and constitutes the d.c. component which is the required output from the phase-sensitive detector. The remaining terms are all oscillatory and, when combined, form the periodic

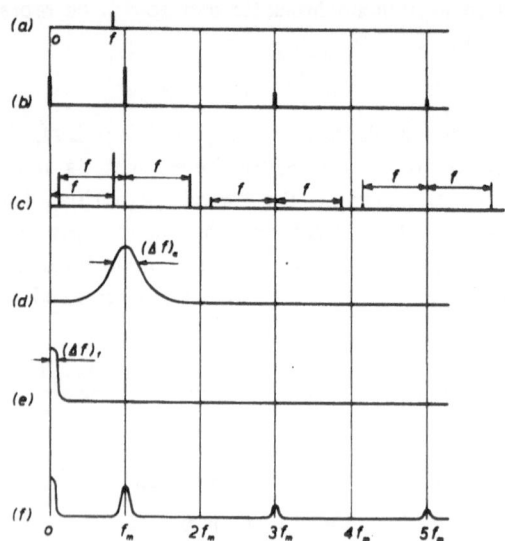

Fig. 2.7. Diagrams showing the noise and signal response
of a double-modulation-recording system. (a) Input signal.
(b) Switching terms. (c) Output signal fed to low-pass filter.
(d) Frequency response of narrow-band amplifier. (e) Frequency response of low-pass filter. (d) Effective input frequency response of phase-sensitive detector with output filter.

fluctuation about the mean d.c. output that is shown is Fig. 2.6.
The fluctuation can be removed by a low-pass filter. Strictly, one
should write the input signal as $v_{in} \sin(\omega t + \phi_p)$, where ϕ_p is
the phase angle between the input signal and the switching func-
tion $h(t)$. The d.c. output term becomes $v_{in}/\pi . \cos \phi_p$ and so
the detector is truly phase-sensitive. If the phase shifter is adjusted properly, the angle ϕ_p is zero, so the term $\cos \phi_p$ is
equal to unity and can be omitted. As previously discussed (p.25),
the frequency of the input signal f is also subject to a small
spread, which is transferred to the nominally-zero-frequency component at the output. Thus, if the system is to pass all of the
signal, the bandwidths of the narrow-band amplifier and of the output filter should not be smaller than the frequency spread on the
input signal, that is, not smaller than the inverse of the time

taken to trace out a line of the spectrum. Applying this criterion to the output filter is slightly tricky, because the frequency spread carries over for half of its extent into the region of negative frequency. However, negative and positive frequencies are physically indistinguishable so that, as shown before, the required cut-off frequency is just half of the bandwidth required for the narrow-band amplifier. To minimize the effects of noise, the system bandwidth must be no greater than that needed to avoid distortion of the signal. To achieve a suitably low system bandwidth, it is not necessary for the narrow-band amplifier and the output filter to be equally low. One or the other will suffice. For example, if the bandwidth $(\Delta f)_a$ of the narrow-band amplifier is greater than that of the output filter $(\Delta f)_f$, as indicated in Figs. 2.7(d) and (e), then any noise component from the amplifier of frequency f, which is separated from the modulating frequency f_m by more than the bandwidth of the output filter, will be translated to a frequency $f_m - f$, which is beyond the band-pass of the output filter, and will not be recorded. This state of affairs is somewhat fortunate because it is generally much easier to design a low-pass filter with a low cut-off frequency than to design a highly selective narrow-band amplifier. For example, the typical bandwidth requirement of 1 c/s can be obtained by using the simple circuit of Fig. 2.5(b) and making $R_1 = 100$ kΩ and $C_1 = 1 \cdot 6$ μF. These values are obtained by picking the convenient value of 100 kΩ for R_1 and calculating the value of C_1 from equation (2.7). In contrast, the normal value of the modulating frequency is 100 kc/s and to design a 100 kc/s amplifier with a bandwidth of 1 c/s would be almost impossible.

Whilst it is not particularly critical how much the bandwidth of the narrow-band amplifier is in excess of that of the final d.c. amplifier, it is inadvisable to try to use a fully wide band amplifier. If the frequency f of the input is arranged to be any odd multiple of the modulating frequency, or indeed made equal to zero, then, from equation (2.11), or from Fig. 2.7(c), it will be clear that a zero-frequency output term is formed between the input signal and the component of the switching series of the same value. This d.c. or near d.c. output signal then falls within the response of the output filter and is recorded. Thus the phase-sensitive detector and output filter alone not only have an effective input response centered at the modulating frequency and extending by the bandwidth of the output filter on either side, but also have spurious responses of the same width at every odd harmonic of the modulating frequency, and also at zero frequency. The complete effective input response is therefore that shown in Fig. 2.7(f), which also indicates how the strength of the spurious responses decay as the amplitude of the switching term from which they stem decreases with frequency. It is therefore necessary, if these spurious responses are not to contribute additional noise, to ensure that the band-pass of the narrow-band amplifier is narrow enough to exclude them. Therefore, according to the above argument, the response of the narrow-band amplifier should not range beyond values between $(\Delta f)_f$ and $3f_m - (\Delta f)_f$.

It is possible that spurious responses may be obtained at the even harmonics of the modulating frequency, since their absence is due to the switching series having no even order terms. These even terms are only absent if the switching function is completely symmetrical and has a 'mark-to-space' ratio of unity. It is therefore advisable to restrict the response of the narrow-band amplifier to values well above $(\Delta f)_f$ and well below $2f_m - (\Delta f)_f$. A practical design along these lines is considered in Chapter 8.

§2.4

THE BLOCH MODEL

In order to progress further with the design of the microwave section of an E.S.R. spectrometer, it is necessary to describe briefly the model of magnetic resonance developed by Bloch.[1] Bloch's model gives a more detailed idea of the nature of the interaction between a spinning electron and a magnetic field than has been presented so far, and although originally devised to explain nuclear magnetic resonance (N.M.R.), it is equally applicable to electron resonance. There are, however, a few minor differences. The charge on a nucleus is positive, whilst that on an electron is negative. The magnetic moment of the nucleus is far less than that of the electron, so that, for equivalent applied fields, the energy splitting, and therefore the resonant frequency, is less for the nucleus. Thus, whilst the electron resonance is normally in the microwave region, the nuclear resonance is in the radio-frequency region.

Bloch likened the spinning nucleus to a child's top [Fig. 2.8(a)]. If the top is set at an angle to the vertical, it is well known that the effect of the torque exerted by the gravitational field is not to make the top tip over, but to induce a uniform precession about the vertical. Similarly, the nucleus comprises a spinning mass and, because of its magnetic moment, experiences a torque when placed in a magnetic field. Thus, even when considered classically, the nucleus does not respond to the applied field by altering the angle between its direction of spin and the field and thereby changing the energy of the system; it remains

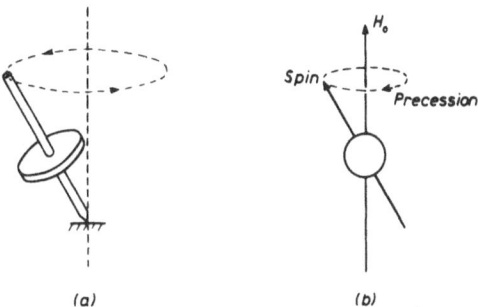

(a)   (b)

Fig. 2.8. Comparison between a spinning top (*a*) and a nuclear spin (*b*) in a steady magnetic field.

at a fixed angle, precessing round the applied field at a constant rate. Consequently, one has even here the idea of a fixed energy state. The main point of interest is that when the frequency of precession is calculated from the classical equations of motion, it is found to agree with the frequency for magnetic resonance given by the nuclear counterpart of equation (1.5). Carrying the analogy a little further, it is also known that a method of reducing the angle between the simple top of Fig. 2.8(a) and the vertical, and therefore of feeding energy into the system, is to apply a force to the end of the top in the direction of the locus of its precessional motion. The end of the top then spirals inwards. Fundamentally, the effect of the force is to apply a torque to the top in the plane of precession. The appropriate way of applying a similar torque to the nucleus is to provide in a plane perpendicular to the steady magnetic field a rotating magnetic field of the appropriate frequency and phased to maintain a direction 90° ahead of the component of magnetic moment in this plane. Experiment confirms that these are the optimum conditions for observing nuclear and electron resonance, and the analogy is therefore correct.

The following additional important points must thus be taken into consideration when designing an E.S.R. spectrometer. Firstly, it is the magnetic, rather than the electric, component of the electromagnetic radiation that is operative in inducing the absorption of photons of energy from the radiation. Secondly, the radiation must be circularly polarized, and, thirdly, the plane of polarization must be at right angles to the steady magnetic field.

When translating these requirements into practice it is quite easy to satisfy the first and last requirements, but, because the fields within a normal microwave cavity are invariably polarized linearly, it is rather difficult to provide circular polarization of the microwave field. Fortunately, a linearly polarized field is almost as effective; for any linearly polarized, alternating field can be resolved into two circularly polarized components rotating in opposite directions. The spin system responds to the one rotating in the right direction, and regards the other as being in error by twice the precessional frequency. The pattern of electric and magnetic fields in a microwave cavity suitable for the E.S.R. spectrometer shown in Fig. 1.4 is given in Fig. 2.9. Here, as in all standing wave systems, the electric and magnetic field patterns are well defined and the nodes of electric field coincide with the antinodes of magnetic field and vice versa. The sample is placed in one of the regions of magnetic field and the cavity is so arranged that the direction of the magnetic field is at right angles to that provided by the magnet.

The above arguments, although giving results which agree with practice, are not particularly rigorous, since a classical model has been used to describe a system whose scale is such that it properly requires a quantum-mechanical description. Slichter[2] has given a rigorous treatment in which he shows that the expectation value of the nuclear magnetic moment follows the

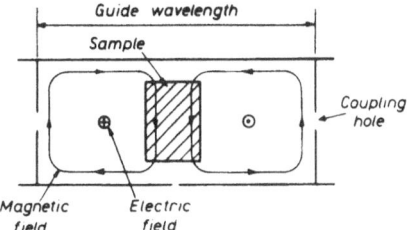

Fig. 2.9.   Electric and magnetic field configurations in a
rectangular E.S.R. cavity.

classical equations of motion exactly, whilst the instantaneous
value is subject to the normal uncertainty. Thus the bulk magnetization of a sample containing many spins will follow the
classical equations, since bulk behaviour is determined by the
expectation value. In order to give a complete picture it is important to include the effects of relaxation mechanisms, because
they, amongst other things, help to explain a further point that
has so far been largely evaded.

Let us return to the simple top of Fig. 2.8(*a*). It is possible to
extract energy from the top by applying the rotating force in the
opposite direction to the direction of precession, whilst maintaining the same sense of rotation of the point of application.
The top then does work against the origin of the force and spirals
out and down. If the nucleus is to absorb rather than emit radiation, the rotating magnetic field must be correctly phased.

The effects of the relaxation mechanisms are twofold. Firstly,
the transverse relaxation processes tend to maintain a random
distribution of the phases of the rotating components of magnetic
moment of the various nuclei. Consequently, any bulk component
of rotating magnetization induced by the phasing together of a
group of nuclei is dispersed after a time, T_2, known as the
transverse relaxation time. Secondly, the longitudinal relaxation
processes tend to restore the Boltzmann distribution between the
two energy states where more spins are aligned with the field
than against it. Therefore, before the rotating magnetic field is
applied, there is no net rotating component of magnetization, and
only a small static component due to the excess population of
spins in the lower energy state, is aligned with the applied
steady field. As there is no initial precession of the magnetiza-

tion, the question of correct phasing does not arise. Furthermore, when a rotating magnetic field is applied, if precession is to occur at all, it must be with an increase in the angle between the spin system and the steady field. Hence the energy of the spin system must be increased, and, for reasons of conservation of energy, the resultant phase relationship of the rotating component of magnetization must be that for which the field does work on the spin system and not vice versa. In fact, when the situation is viewed from a frame of reference which rotates with the rotating field, Pake[3] has shown, by suitably modifying the equations of motion, that the magnetization then turns slowly around the direction of the now-fixed rotating magnetic field in a plane perpendicular to it. This motion corresponds to the increasing angle of precession in the normal frame of reference and indicates that the rotating component of magnetization due to the precession is at right angles to the direction of the rotating field. Furthermore, the direction of the rotating magnetization is in fact 90° behind the field, which is just the position for the maximum torque to be exerted on the spin system by the field in the direction of rotation. Thus the spin system is pulled round by the field and work is done on the spins by the field as expected.

A further function of the relaxation mechanisms is to determine the linewidth of the magnetic resonance. Suppose for the moment that the process described above were repeated with a rotating field of a frequency slightly lower than the precessional frequency of the magnetization. Initially the magnetization would be tipped out from its original position of alignment with the steady field to precess so that the rotating component of magnetization would be 90° behind the rotating field. Work would then be done by the rotating field on the spin system. After a short while, the angle of 90° would be reduced as the magnetization caught up with the rotating field. At the moment when the angle is reduced to zero there would be no torque between the spin system and the rotating field, and therefore no transfer of energy. The angle between the precessing magnetization and the steady field would then remain constant. Subsequently, the magnetization would overtake the rotating field and then the torque would be against the motion of magnetization. Energy would then be transferred

back to the rotating field, and the angle between the magnetization and the steady field would subside back towards zero.
Exact parallelism of the magnetization with the steady applied
field would be restored when the magnetization had reached a
position 90° ahead of the rotating field. A situation similar to the
initial condition then prevails and the process repeats, the overall effect being that energy is transferred to and fro between the
rotating field and the spin system, with no net transfer of energy
in either direction. The situation is modified, however, by the
transverse relaxation mechanism. If the transverse relaxation
mechanism can destroy the rotating magnetization by dephasing
its components before the magnetization catches up with the
rotating field, then the energy previously absorbed from the field
is never returned, and energy flows continuously from the field
to the spin system.

The periodic interchange of energy between the spin system
and the rotating field is at the difference between the frequency
of rotation of the field and the natural frequency of precession of
the spin. Thus a steady transfer of energy is maintained so long
as the frequency difference is less than $1/T_2$. The width of the
magnetic resonance is therefore determined by T_2 and is strictly
equal to $1/\pi T_2$.

The effect of the flow of energy on the spin system is initially
to increase its effective or 'spin' temperature. One can appreciate
this temperature rise by noting that the effect of tipping out
the original static component of magnetization to make it precess
is partly to produce a rotating component of magnetization and
partly to reduce the magnitude of the component in the original
direction. When the dephasing effects have eliminated the
rotating component, the reduced static component remains. Reverting to the two-energy-level model of the system, the original
static component was seen to be due to the excess population of
spins in the lower energy level, each of the spins having a
steady component in line with the field from the magnet. The excess has thus been reduced which means, according to equation
(1.6), that the temperature of the system has been increased. As
the effective 'spin-temperature' increases above that of the
lattice, the spin-lattice relaxation mechanisms then induce the
energy to flow from the spin system to the lattice, so checking

the temperature rise.

Another consequence of the relaxation processes is that, for the condition described, the net magnetic moment will precess behind the rotating field at an angle which, on the average, is rather less than 90°, since the relaxation process is not immediate, and some degree of 'catching up' occurs. For rotational frequencies well below the precessional frequency, the catching up is nearly complete and the angle tends to zero. Conversely, for

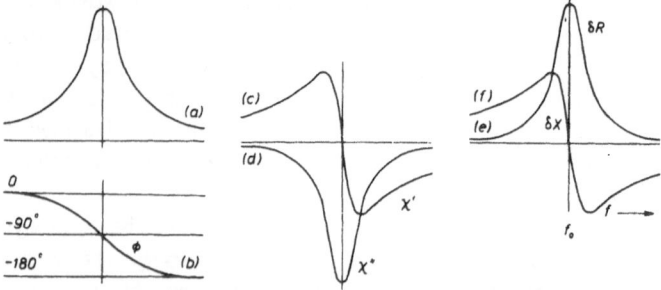

Fig. 2.10. Variation of (a) amplitude of rotating magnetization of sample, (b) phase of rotating magnetization relative to applied rotating field, (c) in-phase component of susceptibility χ'_+, (d) quadrature component of susceptibility χ''_+, (e) resistance δR reflected into a coil by the sample, and (f) reactance δX reflected into a coil by the sample, as functions of the rotational frequency f of the applied field relative to the resonant frequency of the sample f_0.

rotational frequencies in excess of the precessional frequency, the angle by which the magnetic moment lags the rotating field increases to a value approaching 180°. The above results are summarized in Figs. 2.10(a) and (b), which show the amplitude of the net rotating magnetic moment and its phase relative to the rotating field as a function of the frequency of the rotating field. Often the component rotating magnetization in the direction of the rotating field is referred to as the 'in-phase' component and the component leading the rotating field as the 'quadrature' component. Moreover, for rotating fields of low amplitude, the amplitude of the rotating magnetic moment is proportional to that of the applied field. The constant of proportionality is termed the 'rotational susceptibility', χ_+, and if χ_+ is expressed as the complex number $\chi'_+ + i\chi''_+$, χ'_+ represents the in-phase component of the susceptibility and χ_+'' the quadrature component. Curves of χ_+' and χ_+'' are shown in Figs. 2.10(c) and (d). As absorption is proportional to the component of magnetic moment *lagging* the rotating field, the absorption is proportional to $-\chi_+''$ and so positive at resonance. The in-phase term, χ_+', gives no torque, and does not transfer power, and so is not observed in the ordinary way.

A proper mathematical treatment of the above can be found in many of the standard texts on magnetic resonance. The value of χ_+ is given by the expression[4]

$$\chi_+ = -\frac{j2\pi T_2 g\beta M_z}{h[1 + jT_2(\delta\omega)]} \tag{2.12}$$

where M_z is the component of bulk magnetization in the direction of the field provided by the magnet and $(\delta\omega)$ the difference between the angular frequency of the applied a.c. field and the frequency of magnetic resonance. For linearly polarized fields, it is more convenient to define a linear susceptibility, χ, which relates the value of the applied field to the linear component of the rotating magnetization in the direction of the applied microwave field. It then becomes a simple matter to show that $\chi = \chi_+/2$. Hence it is possible to split equation (2.12) into its real and imaginary parts to obtain χ' and χ'', the real and imaginary parts of χ. Thus

$$\chi' = \frac{-\pi T_2^2 g\beta M_z(\delta\omega)}{h[1 + T_2^2(\delta\omega)^2]} \tag{2.13}$$

and

$$\chi'' = \frac{-\pi T_2 g\beta M_z}{h[1 + T_2^2(\delta\omega)^2]} \tag{2.14}$$

and these confirm the general shapes of Figs. 2.10(a − d) and also show that the linewidth is inversely proportional to T_2, as previously suggested.

§2.5

CIRCUIT MODEL OF SPECTROMETER

To obtain an equivalent circuit for the microwave section of an E.S.R. spectrometer it is necessary to consider the basic circuit used for observing N.M.R. Generally, the method for observing N.M.R. is to place the sample in a coil and to pass through the coil a current at the frequency for nuclear resonance. The sample experiences a linearly polarized, sinusoidally varying magnetic field, which must be arranged to be at right angles to the field provided by the magnet. The appropriate rotating component of the linearly polarized field then sets up a rotating component of magnetization within the sample. The magnetization generates a rotating magnetic flux, which induces an e.m.f. in the coil. This e.m.f. is proportional to the rate of change of flux and, therefore, to the rate of change of magnetization, and leads the magnetization by 90°. The in-phase component of susceptibility χ' produces a voltage that leads the current, and the quadrature component χ'' a voltage in antiphase with the current. The effect of χ' is to introduce a further effective inductance into the coil, and that of χ'' to introduce a negative resistance. The corresponding values, δR and δX, are given in Figs. 2.10(e) and (f) for comparison with values of χ'' and χ'. Because χ'' is always negative, δR is positive and behaves as a normal resistance. Thus the effect of power absorption is as if a small resistance had been inserted into the coil, whilst an additional reactive effect, known as dispersion, is introduced and shown to result in a small change in the inductance of the coil. Furthermore, the composite effect has been considered as originating from a rotating magnetization located in a coil. This model is very much akin to an electrical generator and is useful in deciding the best arrangement for a spectrometer.

The main objective is to couple as efficiently as possible to the detector the output from the spin-generator formed by the sample and the coil. In other words, the generator must be matched to the load presented by the detector. A convenient method of doing this is to resonate the inductance of the coil with a suitable capacitor C and then to couple inductively the detector to the coil, as in Fig. 2.11(a). The generator is then presented with a zero-reactance impedance and any resistive impedance that is reflected into the resonant circuit by the detector coupling. As the resistance of the coil is the effective output impedance of the generator, maximum transfer of power will occur when the coupling is such that the reflected load impedance $(\omega^2 M^2)/R_0$ is equal to the resistance of the coil. In addition, short of saturation, the amplitude of the spin-generator voltage is proportional to the current driven through the coil by the main a.c. generator. For

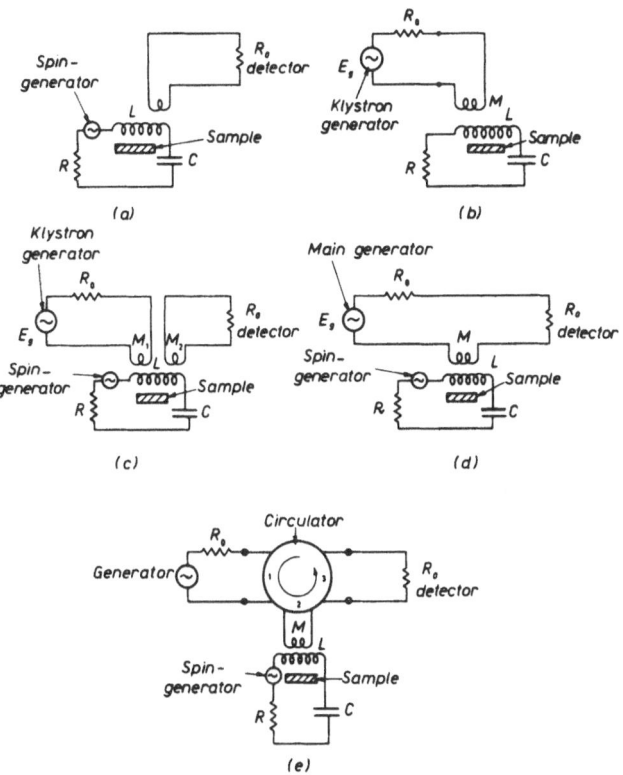

Fig. 2.11. (a) Ideal method of coupling signal from sample spin-generator to detector. (b) Ideal method of coupling klystron power to cavity. (c) Transmission spectrometer. (d) Absorption spectrometer. (e) Reflection spectrometer with a circulator.

such a current to flow, power is required because the coil has a resistance. Ideally, all of the available power from the main generator should be coupled into the coil. The appropriate arrangement is shown in Fig. 2.11(b). Again the coupling should be such that the load reflected by the main generator into the coil is equal to the resistance of the coil. These two requirements are clearly in conflict and a compromise has to be sought, such

as that of Fig. 2.11(*c*) or (*d*). Both arrangements are less than
ideal, because some of the main-generator power is fed to the
detector and some of the spin-generator power to the main generator. It is fortunate, therefore, that a device known as a 'circulator'
exists. A circulator is a multi-port component which ensures
that a signal fed in at one of its ports comes out at the next
adjacent port only. For the arrangement of Fig. 2.11(*e*), the main
generator is matched to its associated transmission line which
thus delivers all of the available power from the generator to the
first port of the circulator. This power leaves at port 2 and
propagates towards the resonant circuit which, because it is
matched to the line, absorbs all of the power; none is reflected.
Then, because the resonant circuit is matched to the line, all of
the available power from the sample spin-generator is propagated
to the circulator and reaches the detector from port 3. All the
available klystron power is therefore dissipated in the resonant
circuit and all the available spin-generator power reaches the
detector. This arrangement is ideal and one is merely left with
the problem of designing a cavity which gives the greatest available power from the spin-generator for a given amount of power
absorbed from the main generator. This point is discussed in
Chapter 4.

The compromise circuit of Fig. 2.11(*c*) is analogous to the
E.S.R. spectrometer circuit of Fig. 1.4. The generator is equivalent to the klystron and the section of waveguide to which it is
connected and matched, the resonant circuit to the cavity resonator, and the load to the microwave detector and the section of
waveguide to which it is connected and matched. The mutual
inductances correspond to the coupling holes in the irises. It
thus becomes apparent that this arrangement, known as the transmission-cavity spectrometer, can be improved by using the microwave equivalent of Fig. 2.11(*e*). Fortunately, microwave circulators do exist, so the corresponding arrangement of Fig. 2.12 is
quite feasible.

§2.6

MICROWAVE DRIVE PHASE RELATIONSHIP

An important factor which determines how well the detector
responds to the signal transmitted to it from the sample is the

phase relationship between the sample signal and the signal originating from the klystron generator. The effect is best under-

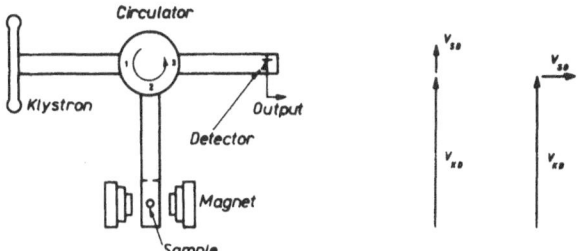

Fig. 2.12. E.S.R. reflection-cavity spectrometer with a circulator.

Fig. 2.13. In-phase and quadrature phase relationships of microwave signal voltage V_{SD} and microwave voltage V_{KD} from the klystron at the detector.

stood by using vectors to represent the two components of microwave voltage at the detector, as in Fig. 2.13. Here V_{KD} represents the voltage at the detector due to the klystron and V_{SD} the voltage at the detector due to the sample. Normally the klystron voltage is much larger than the sample voltage so that, when the two voltages are in phase, as in (*a*), the change in the amplitude of the resultant is equal to the amplitude of the signal voltage, and, when the two are in quadrature, the change in amplitude due to the sample voltage is, to the first order, zero. In fact, this is the reason why dispersion is not commonly observed in an E.S.R. spectrometer. In §2.5 it was shown that the dispersive and absorptive voltages induced by the sample are in phase quadrature, and normally the arrangement of the system is such that the absorptive component produces at the detector a voltage which is actually in antiphase with that from the klystron. Thus the effect of the absorptive component is to produce a first-order reduction in the detected signal level, whilst the dispersive component of voltage arrives at the detector in quadrature with the klystron voltage and so is suppressed.

A typical example of this situation occurs in the transmission-cavity spectrometer of Fig. 1.4. If the equivalent circuit of this arrangement, shown in Fig. 2.11(*c*), is redrawn with the load and generator shown reflected into the cavity circuit, as in Fig.

2.14, then it is obvious that, provided the circuit is at resonance, the current flowing from either the absorptive or dispersive components of voltage induced by the sample will be in phase with their respective voltages. However, the absorptive component of voltage is in antiphase with the current due to the klystron, whilst the dispersive voltage is in quadrature. Consequently, the absorptive current through the detector is in antiphase with the current due to the klystron and results in a first-order reduction. The dispersive current, on the other hand, is in quadrature and is suppressed. Here the arguments previously applied to voltage vectors have been used for current vectors; this is permissible. If this argument is not clear, one can translate the currents to voltages via the detector resistance into which they flow.

Another way of considering the above effect is to recall that the absorptive component of the sample voltage is equivalent to a small additional resistance placed in the circuit and the dispersive component to a small reactance. If, now, the current flowing through the detector in Fig. 2.14 is plotted against the circuit reactance, a response curve similar to Fig. 1.9 will be obtained. Thus, when the circuit is at resonance, the slope of the curve which determines the sensitivity to a small change in reactance is zero, and so only a second-order response to the sample reactance is obtained. However, a change in the circuit resistance is observed without suppression at resonance, and, in fact, has its greatest effect there, since the total circuit impedance is then at its smallest compared with the sample resistance. If, however, the circuit is detuned, then the slope of the cavity response curve becomes finite and the dispersive and absorptive effects become mixed. Similarly, in the previous argument, if the circuit is not at resonance, the phase relationships between current and voltage are disturbed and the dispersive sample current is no longer in exact quadrature with the current due to the klystron.

The vector model of Fig. 2.13 can also be used to examine the effects of the phase relationship in the reflection-cavity spectrometer of Fig. 2.12. For optimum sensitivity, the cavity is exactly on tune and matched to the waveguide, so that no signal from the klystron reaches the detector by reflection from the cavity. Consequently, the only signal to reach the detector is that from the sample and therefore suppression does not occur, and so both absorption and dispersion can be observed. It is

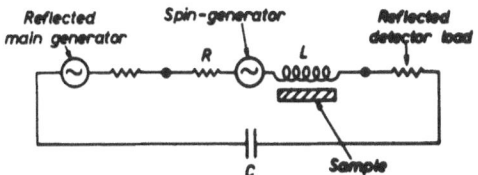

Fig. 2.14. Equivalent circuit for a transmission-cavity spectrometer shown with the load and klystron generator reflected into the cavity circuit.

sometimes erroneously supposed that the shape of the resonance line then observed is a simple superposition of the individual absorption and dispersion line shapes shown in Figs. 2.10(*e*) and (*f*). This is not so because the signals must be combined vectorially since they are orthogonal in phase. Fig. 2.15 shows the relative phases of the absorption and dispersion signals, together with their variations of amplitude with frequency. These combine to give a resultant which actually describes a circle. Thus the amplitude of the vector sum of the two signals is a curve with a shape somewhat similar to that for the absorption line. The form of the curve is not exactly the same however,

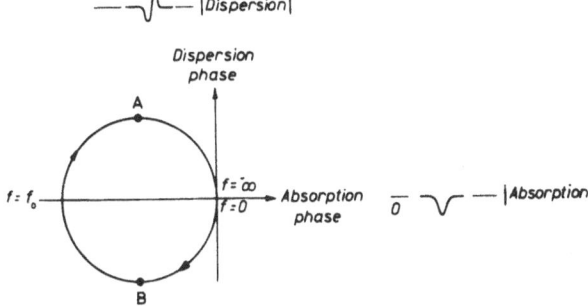

Fig. 2.15. Vector diagram showing combination of absorption and dispersion signals from sample.

and, in particular, the width is slightly increased. The points A and B correspond to the width at half-height for the absorption

signal, but, at these frequencies, the height of the combined
signal is only reduced by $1/\sqrt{2}$ from its maximum. In fact, the

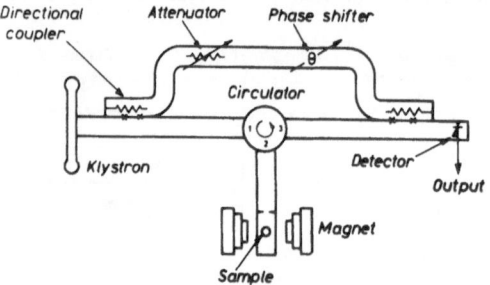

Fig. 2.16. Reflection-cavity system with a bucking arm.

combined signal is proportional to the original rotating magnetization of the sample, which is given in phase and magnitude by
Figs. 2.10(*a*) and (*b*). Thus a close inspection of Figs. 2.10(*a*)
and (*d*) will confirm that (*a*) is slightly wider. The circle of
Fig. 2.15 also shows clearly how the phase of the magnetic
moment swings through 180° as the frequency passes from one
side of resonance to the other.

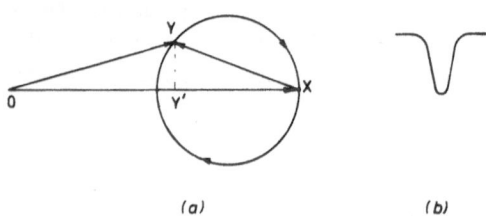

(*a*)   (*b*)

Fig. 2.17. (*a*) Vector addition of bucking OX and sample XY
signals for a reflection-cavity spectrometer with bucking.
(*b*) Variation of resultant OY.

If a reflection-cavity spectrometer is required to provide information about absorption and dispersion separately, a 'bucking
arm' should be included in the system, as in Fig. 2.16. By this
means, a steady, additional signal can be fed from the klystron
to the detector, and the phase and magnitude of the bucking
signal can be adjusted by the phase shifter and attenuator to
give whatever degree of suppression required. The effect of the
bucking signal is illustrated in Fig. 2.17, where OX represents

the bucking signal, to which is added the signal from the sample, XY. The extremity, Y, of the sample signal starts at X to give a zero signal when the frequency is well below resonance, and traverses the circle in the anti-clockwise direction to reach X again when the frequency has passed through and well beyond resonance. The resultant which reaches the detector is therefore OY which, provided OX $\gg$ XY, varies in amplitude in a manner almost identical with OY', that is, with the absorptive component XY' of XY. The dispersive component YY' is largely suppressed. The variation of OY is shown in Fig. 2.17(*b*). Fig. 2.17 is drawn for the normal phase relationship between signal and bucking components where the intention is to suppress dispersion. The effects of using this and the three other principle phase relation-

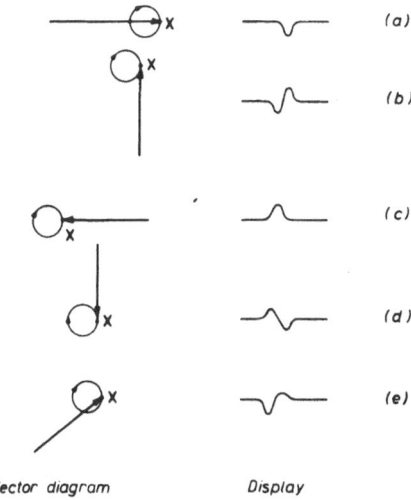

Vector diagram   Display

Fig. 2.18. Effect of altering the phase of the bucking signal in a reflection-cavity spectrometer upon the observation of absorption and dispersion.

ships are shown in Figs. 2.18(*a* – *d*). Thus in (*b*) and (*d*) the phase relationship is shifted by $\pm 90°$ to give suppression of absorption and observation of dispersion, whilst (*c*) simply reverses the sign of the absorption signal; (*e*) is an intermediate phase relationship which mixes both absorption and dispersion.

§ 2.7

MICROWAVE DRIVE AMPLITUDE

In addition to the need for a bucking signal to resolve the effects of absorption and dispersion in a reflection-cavity spectrometer, some steady level of microwave drive is required at the detector to obtain an adequate detector efficiency.[5] This point is best illustrated by the curves of Fig. 2.19 which represent the operation of the detector of a reflection-cavity spectrometer with two different levels of bucking signal and with the phase relationship arranged to select the absorption component of the sample signal.

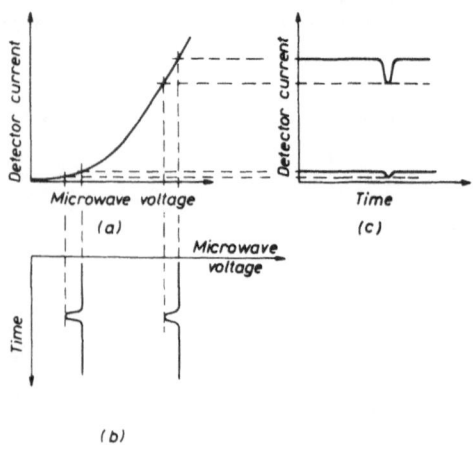

Fig. 2.19. Variation of detector efficiency with bucking level. (*a*) Detector characteristic. (*b*) Variation of microwave signal at detector for two levels of bucking. (*c*) Resulting variation in detector current.

Here (*a*) represents a simple plot of detector output current against microwave input voltage, whilst (*b*) represents the variation of microwave voltage with time when the resonance is traversed. This variation is independent of the amplitude of the bucking signal and is equal to the sample voltage. The resulting variation in detected output current is then roughly proportional to the slope of curve (*a*) at the operating point, and this slope is dependent on the amplitude of the bucking signal. In fact, the slope of the detector characteristic (*a*) is almost linear for high levels of input power (> 1 mW), but becomes curved and eventually parabolic for lower levels. Thus to obtain the maximum detector efficiency, which is necessary if the detected signal is to be as large as possible compared with the noise from the amplifier, the amplitude of the bucking signal must be large enough to approach the linear region of the characteristic. If no bucking is used at all, the slope of the characteristic becomes zero and the efficiency of the detector, to a first approximation, becomes zero.

The form of detector characteristic of Fig. 2.19(a) is largely determined by the static characteristic of the diode, that is, the plot of d.c. current against applied d.c. voltage for the diode itself. As the detector is generally a normal point-contact semiconductor diode, the form of the curve is similar to the example shown in Fig. 2.20. If a sine-wave voltage generator is connected directly across such a diode, the current waveform depends, to a great extent, on the amplitude of the input voltage. For very low input voltages, as in Fig. 2.20(a), the characteristic approximates to a straight line and the output current is nearly sinusoidal. Thus the instantaneous output current is as often negative going as positive, and so no mean d.c. output current is obtained. For an increased amplitude of input voltage, as in (b), the curvature begins to become significant, so that the output current is distorted to give larger positive-going peaks of current than negative-going peaks. Thus a steady component of output current begins to flow. For quite a large range of rather greater amplitudes of input voltage, the detector characteristic does not differ greatly from the idealized characteristic described by lines AB and BC. The curve of the output current then takes the form shown in (c′) and is strictly proportional to the amplitude of input voltage. Fig. 2.20 (c′) corresponds to the region of constant slope in Fig. 2.19, whilst (a) and (b) correspond, for very low input levels, to near zero detector efficiency and, for the slightly higher levels of (b) and (b′), to the region of curvature.

The curved region of Fig. 2.19 is sometimes termed the region of 'square-law' operation, whilst the region of constant slope is said to correspond to 'linear' detection. In my experience the square-law region is rarely anything like parabolic until levels of drive far below those normally used are reached. There is, in fact, a fairly wide region where the dependence can become almost quartic and certainly cubic. Thus these terms should be viewed with caution.

The above description is a little over simplified. For practi-

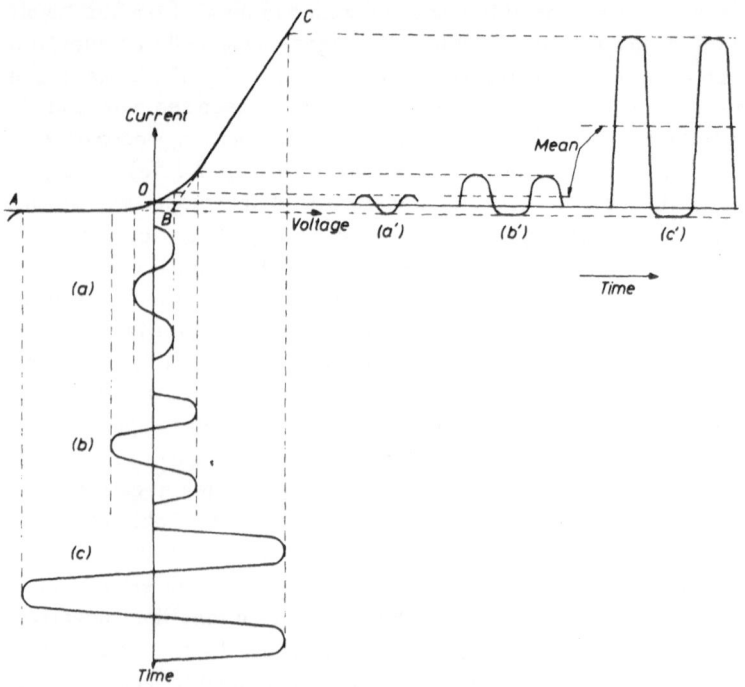

Fig. 2.20. Use of a static detector characteristic to convert various levels of input voltage into output current.

cal systems it is necessary to connect some sort of load between the detector and the microwave 'a.c. generator'. For crystal-video spectrometers this load may be a resistor or possibly a transformer. For double-modulation spectrometers, which are the most sensitive E.S.R. spectrometers and therefore of most interest, the transformer is likely to be tuned. When a load is connected, both d.c. and a.c. voltages develop across the load and must be taken into consideration. Consequently, the above arguments become no longer quantitatively correct and can only be used to give a general indication of the results to be expected. However, results obtained with practical detecting systems confirm the general accuracy of the approach.

The other factor to be considered when determining the opti-mum level of microwave drive to the detector is the noise from the detector itself. If amplifier noise were the sole contribution to the overall noise level, then the signal-to-noise-ratio would simply be determined by the variation of detector efficiency with microwave drive, that is, the signal would increase roughly in accordance with the slope of Fig. 2.19(*a*) to follow a curve of the type shown in Fig. 2.21(*a*), whilst the noise remained constant. However, the crystal noise increases with microwave drive, as does the noise contribution from the klystron; therefore the variation of signal-to-noise ratio depends on the individual variations of detector efficiency and noise from the amplifier plus crystal and klystron.

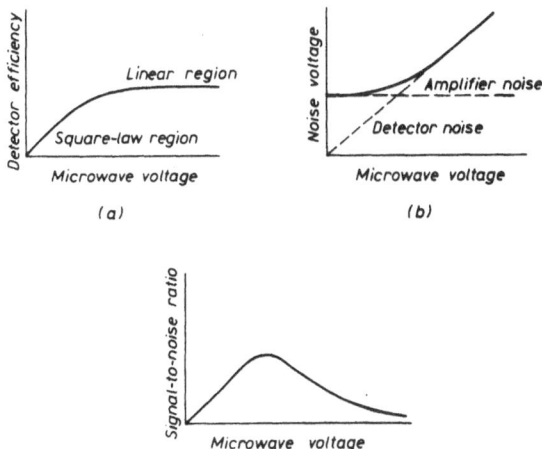

Fig. 2.21. Graphs of (*a*) detector efficiency, (*b*) output noise voltage, and (*c*) signal-to-noise ratio for a spectrometer sys-tem as the microwave bucking level is varied.

Fig. 2.21(*b*) gives a typical curve showing the variation of total noise level and indicates how, at low levels of microwave drive, only amplifier noise is of importance, but, as the drive is in-creased, detector current begins to flow, and shot- and flicker-noise fluctuations in this current are observed. The increase in detector efficiency gives an improved response to the klystron noise, which also increases at the input of the detector, in pro-portion to the klystron power.

In practice it is commonly found that the transition point where crystal and klystron noise become comparable with amplifier noise coincides with the transition between square-law and linear operation of the detector. Thus, for values of microwave drive below the common transition point, the signal-to-noise ratio falls because the noise, being amplifier noise, remains constant, whilst the signal falls with decreasing detector efficiency. On the other hand, for higher values of microwave drive, the noise increases, whilst the detector efficiency levels and gives a constant signal. Thus a change in the level of microwave drive in either direction gives a reduction in signal-to-noise ratio. The transition point between linear and square-law operation of the detector is therefore also the point of optimum microwave drive. These effects are further illustrated by the graph of signal-to-noise ratio shown in Fig. 2.21(c), which was simply obtained by plotting the ratio of the curve of Fig. 2.21(a) to that of Fig. 2.21(b).

So far we have established that the signal-to-noise ratio obtained with the detecting and recording system of an E.S.R. spectrometer depends on the bandwidth of the system, the modulating frequency, and the level of microwave drive to the detector. The dependence of noise on bandwidth is simple, the noise power being proportional to the bandwidth (see §2.2). However, the effects of modulating frequency and microwave drive are not so conveniently expressed, and some experimental determination is needed to check the foregoing remarks. To obtain this information it is neither convenient nor necessary to build a range of spectrometers with varying modulating frequencies and microwave drive levels. In practice the mode of operation of the microwave detecting system of a double-modulation spectrometer is almost exactly identical with that of a double-channel microwave mixer.

Referring back to the vector diagram of Fig. 2.17, the effect of the linear sweep applied during double modulation makes the extremity Y of the sample signal vector XY move slowly round the circle in the direction shown. Sinusoidal field modulation then causes the vector to oscillate along the path of the circle on either side of its mean position. Thus the resultant signal OY reaching the detector is both amplitude-modulated and phase-modulated at the frequency of the field modulation. The microwave detector does not, however, respond to phase modulation and the signal might, to all intents and purposes, only be amplitude-modulated. The form of the amplitude modulation is indicated by Fig. 2.22, where (*a*) represents the variation of the amplitude of the microwave voltage reaching the detector as a function of the applied magnetic field, and (*b*) the applied field modulation for a fixed value of the mean field corresponding to the position of maximum sensitivity, that is, half-way up one side of the resonance line. Provided that the amplitude of sweep is

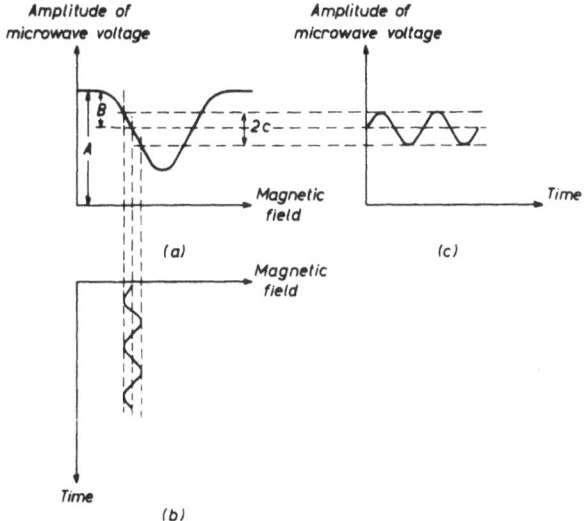

Fig. 2.22. (*a*) Graph showing microwave voltage reaching the detector of a spectrometer as a function of the applied magnetic field. (*b*) Applied field modulation for the position of maximum sensitivity. (*c*) Resulting modulation of microwave detector voltage.

suitably small, the sloping edge of the resonance line translates the field modulation into sinusoidal amplitude modulation at the modulating frequency. Thus, if the phase modulation is neglected, the signal reaching the detector can be written, from the notation of Fig. 2.22, as

$$v_D = (A - B)\cos\omega t + C\cos\omega_m t . \cos\omega t \qquad (2.15)$$

where $\omega = 2\pi f$, f being the microwave frequency, and $\omega_m = 2\pi f_m$, f_m being the modulating frequency. This expression may be expanded to give

$$v_D = (A - B) \cos \omega t + \frac{C}{2} \cos(\omega + \omega_m)t + \frac{C}{2} \cos(\omega - \omega_m)t$$

$$(2.16)$$

where the first term represents the carrier of an amplitude-modulated signal and the second two the upper and lower sidebands. The function of the detector is, therefore, to mix the sidebands with the carrier to produce difference frequency terms equal to the modulating frequency, and so the system is truly a mixer.

Generally, the efficiency of a mixer is expressed as its 'conversion-gain', which is defined as the available output power from the mixer at the difference frequency, compared with the available signal power from the source. It will therefore be clear that the conversion gain is directly proportional to what has so far been termed the detector efficiency.

In considering the microwave detector as a mixer it is also valuable to consider the effect of varying the microwave drive to the detector. It has already been shown that the effect of varying the microwave drive in a spectrometer is simply to alter the overall height A of the curve shown in Fig. 2.22(a) above the base line. The form and amplitude of the curve remains unaltered. Thus, from equation (2.16), the effect upon the output signal is as if the carrier amplitude had been altered whilst the sidebands remain unaltered. Altering the microwave drive power in the spectrometer system is therefore simply equivalent to varying the local oscillator power in the mixer.

Thus, to determine for a double-modulation spectrometer the variation of signal-to-noise ratio with drive power and modulating frequency, all that is required is a microwave modulator capable of giving constant amplitude sidebands, whilst both the modulating frequency and the amplitude of the carrier are varied. Such a system has been devised[5] and is shown in Fig. 2.23. Here the sidebands are obtained by passing a small fraction of the output power from the klystron to an amplitude modulator. The output from the modulator is then recombined with the main carrier which arrives at the detector via a calibrated attenuator. A fixed attenuator is included in the modulator branch to ensure that the carrier emerging from the modulator is always small at the detector compared with that emerging from the calibrated attenuator. The variation of total carrier power reaching the detector is then given accurately by the setting of the variable attenuator, which

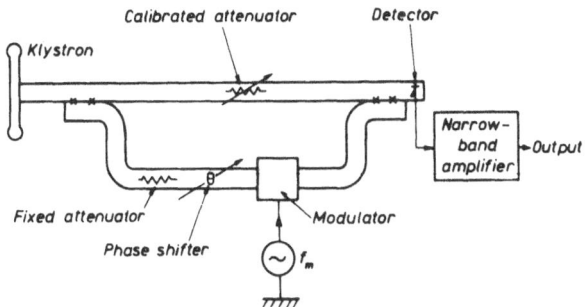

Fig. 2.23. Method of simulating a double-modulation E.S.R. spectrometer to investigate variation of signal-to-noise ratio with microwave drive power

can also be altered without changing the amplitude of the sideband components. The frequency of the generator driving the modulator can also be conveniently varied and the narrow-band amplifier following the detector must be suitably tuned. The phase shifter restores any phase difference between the two paths and ensures that the sidebands are correctly phased relative to the carrier. Using this arrangement, values of signal-to-noise ratio for the detecting system can be obtained as a function of microwave drive for a wide range of modulating frequencies.

The results of these measurements are shown in Fig. 2.24(*a*) and, although labelled 'noise factor', can, for the present, be taken to represent the noise-to-signal ratio. Thus the noise-to-signal curves should show a minimum. This minimum is exhibited with varying degrees of clarity for all values of modulating frequency. Also, the reduction in noise with frequency is clearly shown. The reason for the lack of uniformity in the shape of the curves is best understood by considering the component curves of Fig. 2.24(*b*) and (*c*). These represent the individual variations of signal and noise with microwave drive from which (*a*) the noise-to-signal ratio was obtained. Of these, the noise variation (*c*) is particularly interesting because it shows clearly the transition between amplifier noise on the left-hand side, and crystal

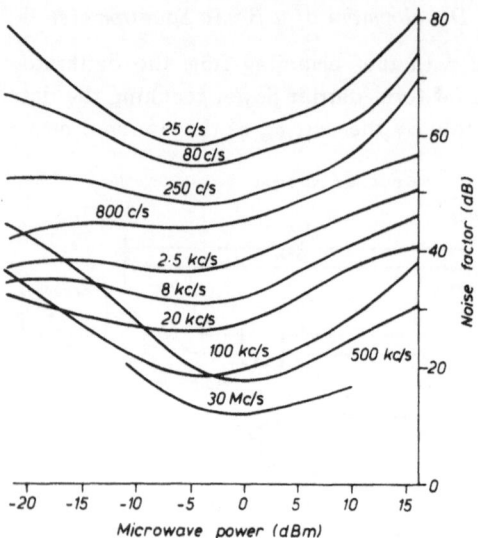

Fig. 2.24(*a*). Variation of the noise factor of the detecting system of an E.S.R. spectrometer as a function of the level of microwave drive to the microwave detector and the frequency of the amplifier following the detector.

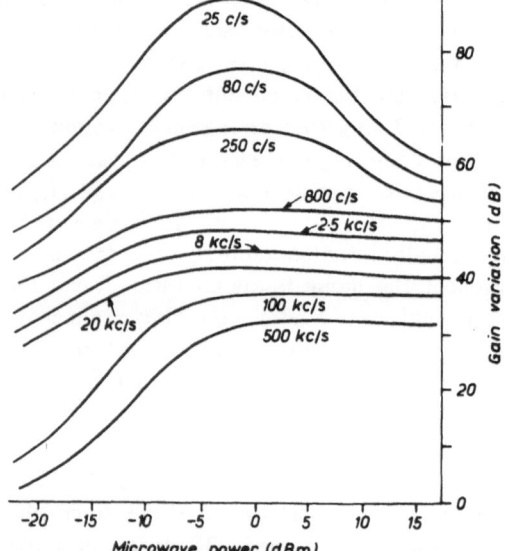

Fig. 2.24(*b*). Curves showing the variation of gain of the microwave detector as a function of microwave drive at various amplifier frequencies.

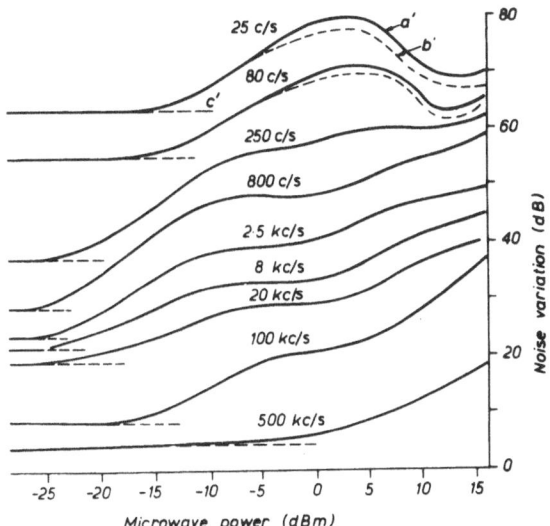

Fig. 2.24(*c*). Variation of overall noise level with microwave drive at various amplifier frequencies. Curve *a'*: overall noise level. Curve *b'*: effect of removing klystron noise. Curve *c'*: amplifier noise contribution.

Note: For (*b*) and (*c*), values of gain and noise are neither absolute nor consistent from one curve to the next.

and klystron noise on the right. Furthermore, a balanced mixer technique can be used to isolate the effects of klystron noise and the results of the test are also shown. Surprisingly, the contribution of klystron noise was almost always negligible, and only at the lowest frequencies and the highest values of microwave drive was any contribution discernible. It should be pointed out, however, that the crystal used was selected as being the average of a batch and that the klystron, an E.M.I. 25157, is generally considered to be very noise-free. It is usual to take some pains in selecting a crystal of above average qualities, so the klystron amplitude-modulated noise may figure more strongly in some systems. In a practical spectrometer there are also effects associated with the cavity that can increase the contribution of noise from the klystron even further. These effects are somewhat complex and are discussed separately in §§ 3.9 and 3.10. The main difference between the observed variation of noise-to-signal ratio shown in Fig. 2.24(a) and that suggested by the foregoing discussion is that whilst for low and high modulating frequencies the expected well-defined optimum occurs at the transition point of microwave drive where the detector efficiency levels off, for intermediate frequencies the optimum becomes much flatter and moves to lower values of microwave drive. The reason for this change is that, as pointed out in §2.2, the $1/f$ noise in a crystal rises above the white noise at amplifier frequencies of below approximately 10 Mc/s, whilst valve flicker noise only becomes significant below approximately 10 kc/s. Hence, for frequencies below 10 kc/s, the ratio of crystal noise to amplifier noise at a given local oscillator power is roughly 1000 times (power ratio) greater than for frequencies above 10 Mc/s. The transition point for the noise curve is thus shifted to a lower value of microwave drive power, whilst that for conversion gain remains constant. Consequently the noise, which at higher frequencies begins to increase at about the same level of drive as the gain flattens off, begins to increase at a point where the gain is still increasing. Furthermore, the increase of noise and gain is about equal and so the signal-to-noise ratio remains roughly constant until the original optimum value of drive power is reached when the gain flattens off. For very low frequencies, the difficulty in obtaining an efficient coupling between the detector and amplifier restores the relative importance of amplifier noise, and thus causes a return to the high-frequency conditions, where the optimum drive again increases and its value becomes more critical. For the intermediate frequencies (250 c/s to 100 kc/s) it is clear from Fig.2.24(c) that, at the optimum value of microwave drive, the contribution of crystal noise is well in excess of the amplifier noise. There is no particular point in attempting to design an amplifier with an exceptionally low noise characteristic for spectrometers operating in this range. On the other hand, for high values of modulating frequency, the crystal and amplifier noise make about equal contributions, so that for normal operation of the double-modulation spectrometer it is important to use a low-noise amplifier.

§2.8

NOISE FACTOR

Some explanation of the term 'noise factor' used in labelling the noise-to-signal-ratio curves of Fig.2.24(a) is now due. Generally speaking, for a measuring system it is important that the noise should be as low as possible. However, one source of noise which is more or less unavoidable is the thermal noise from whatever is being measured, i.e. the source. In the ideal reflection-cavity circulator spectrometer system of Fig. 2.12, the source is the cavity, since the cavity is matched, via the waveguide and circulator, to the detector. The thermal noise radiated to the detector is therefore the same as that from a matched load coupled directly to the detector, and at the temperature of the cavity. Additional noise then originates from the amplifier, detector and klystron, and although this added noise can never be zero, if it can be reduced so that it is small compared with the thermal noise from the source, there is little point in reducing it further. Hence a good criterion for the noise performance of a detecting system is to compare the ratio of the actual observed noise power at the output of the system with the noise power that would be observed if only the thermal contribution from the source were present. This ratio is termed the noise figure F, i.e.

$$F = \frac{\text{Output noise power}}{\text{Output noise power originating from source}} \quad (2.17)$$

The closer F approaches unity, the less point there is in improving further the noise contributed by the detecting system. Often the ratio is expressed in decibels. That is

$$F_{(dB)} = 10 \log F_{(ratio)} \quad (2.18)$$

In all of the above tests the signal level is constant. Thus the ratio of signal to thermal source noise is fixed and, hence, the noise-to-signal ratio is proportional to the noise factor. In fact the thermal noise from a resistance R may be represented by a noise generator of voltage $\bar{v}^2 = 4kTRB$, where k is Boltzmann's constant, T the absolute temperature, and B the bandwidth. The available power from the generator, which has an output impedance of R, is then equal to kTB and independent of R. Thus the output noise power from the spectrometer is given by $FkTBG$, where G is the power gain of the system. The output signal power would similarly be $P_a G$, where P_a is the available signal power from the cavity system. The signal-to-noise ratio is then given by $P_a/FkTB$. Alternatively, if the sensitivity of the detecting system is expressed as the value of signal power $P_{a(min)}$ for which a signal-to-noise ratio of unity is obtained, then

$$P_{a(min)} = FkTB \qquad (2.19)$$

Further evidence for the similarity between the mode of operation of a double-modulation spectrometer and a microwave mixer lies in the well-known expression

$$F = \frac{F_1 + NTR - 1}{G_1} \qquad (2.20)$$

which gives the overall noise figure of a microwave mixer and amplifier in terms of the various factors that determine it. Of these, F_1 is the noise figure of the amplifier, NTR the noise temperature ratio of the mixer, that is the ratio of the available noise power at the output of the mixer compared with the available thermal noise power from a resistor of a value equal to the output impedance of the mixer, and G_1 the conversion gain of the mixer. Thus the numerator of equation (2.20) represents the noise contributions, whilst the denominator determines the output signal level. At very low values of microwave drive the mixer approximates to a resistance of a value equal to the slope of the static characteristic of the diode, as illustrated by Fig.2.20. Thus the NTR is unity so that $F \simeq F_1/G_1$. The noise is therefore constant and determined by the amplifier noise figure, whilst the conversion gain increases with microwave drive, to give an increasing signal-to-noise ratio. Eventually the NTR begins to exceed unity and finally dominates $F_1 - 1$ to make the total noise figure dependent on mixer noise and not at all on amplifier noise. At about the same point, the conversion gain, being strongly linked to the detector efficiency of Fig.2.21(a), ceases to increase and the signal-to-noise ratio decreases simply with the increase in crystal noise. Thus the variation of signal-to-noise ratio with microwave drive follows the form already noted. One point of importance which is emphasized by equation (2.20) is that, since the NTR can never be less than unity, there is never any value in reducing the noise of the amplifier beyond the stage where its individual noise figure approaches unity. Of course, in those instances where the best overall noise figure is obtained at a value of microwave drive which gives an NTR well in excess of unity, there is no point even in trying to obtain an amplifier noise figure of much less than the NTR then prevailing. It is probably worth noting that normal mixer noise figures are quoted for single-channel operation, although they are most commonly double-channel devices of the type discussed. Thus for spectrometer applications it is legitimate to halve the values quoted, or to reduce the figure in decibels by three. Noise figures quoted in Fig. 2.24(a) are expressed for single-channel operation, however. It is clear from the above discussion that both the microwave noise figure. and the amplifier noise figure are of considerable importance. Methods of measuring and improving these quantities are discussed in Chapter 9.

References

1. Bloch, F., *Phys. Rev.*, 1946, **70**, 460.
2. Slichter, C.P., *Principles of Magnetic Resonance* (Harper & Row, 1963).
3. Pake, G.E., *Paramagnetic Resonance* (Benjamin, 1962).
4. Siegman, A.E., *Microwave Solid-State Masers* (McGraw Hill, 1964).
5. Gambling, W.A., Hubble, A.W., and Wilmshurst, T.H., *Measurements on a microwave mixer at intermediate frequencies from 25 c/s to 30 Mc/s* (Paper read at the Symposium on Microwave Applications of Semiconductors, London, 1965).

Chapter 3

Microwave Systems

It will be clear from Chapter 2 that there is considerable flexibility in the way in which the microwave generator, the sample cell and the microwave detector can be interconnected in an E.S.R. spectrometer. Therefore this chapter is devoted to a fairly detailed, quantitative comparison[1,2] of the more commonly used arrangements. We shall develop from the method of analysis used in Chapter 2 a simple graphical method for determining the relative response of any of the systems to the absorptive and dispersive components of the magnetic resonance, and shall compare the sensitivity of the cavity resonator with that of an optimally filled waveguide absorption cell. After considering the question of power saturation of the magnetic resonance, we shall discuss the changes required in a system that is to have optimum sensitivity for a saturable sample, rather than, as up till now, for an unsaturable sample. One or two commonly used compromise systems suitable for both types of sample will then be considered and the extent of the compromise determined.

A factor of considerable importance in choosing a microwave and cavity system for a spectrometer is the ability of the system to prevent noise generated by the klystron from contributing to overall noise level. The above systems are therefore considered from this viewpoint and various modifications tending to improve the rejection of klystron noise are described. Finally some improved methods of recording are discussed.

§3.1

TRANSMISSION-CAVITY SPECTROMETER

In this section we shall determine quantitatively the sensitivity

62

of the transmission-cavity spectrometer of **Fig. 1.4**, using the
equivalent circuit of **Fig. 2.11(c)**. The absorptive component of a
magnetic resonance can be represented by a small resistance
δR, reflected in series with the inductance in the equivalent
circuit of the cavity, and the dispersive component by a reactance δX (see §2.5). The sensitivity to absorption and dispersion
can thus be defined as the change δV_L in detector voltage V_L
for given values of δR and δX, that is, $\partial V_L/\partial R$ and $\partial V_L/\partial X$.
Thus, V_L is determined from the equivalent circuit, and then
the derivatives are obtained. Optimum values for the input and
output couplings, M_1 and M_2, are determined by suitable differen-
tiation, and it can also be shown that the sensitivity is at a
maximum when the cavity is exactly at resonance. Inserting the
values thus obtained into the expressions for $\partial V_L/\partial R$ and
$\partial V_L/\partial X$ we obtain the optimum sensitivities. A similar procedure can then be applied to the reflection- and absorption-cavity
systems of **Fig. 2.11(e)** and (d) to obtain an overall comparison.
From the equivalent circuit of **Fig. 2.11(c)** the detector voltage
V_L in a transmission-cavity spectrometer is given by the expression

$$V_L = E_g \frac{\dfrac{\omega^2 M_1 M_2}{R_0}}{R + jX + \dfrac{\omega^2 M_1^2}{R_0} + \dfrac{\omega^2 M_2^2}{R_0}} \tag{3.1}$$

where E_g is the effective e.m.f. of the klystron, R_0 is the resistive impedance of the source and the detector, which are assumed
to be matched to the waveguide, M_1, M_2 are the mutual inductance
coefficients for the input and output couplings, R and X represent, respectively, the resistive losses and the reactance of the
cavity, and ω is the angular frequency of the microwave source.
The sensitivities $\partial V_L/\partial R$ and $\partial V_L/\partial X$ to absorption and dispersion then become respectively

$$\frac{\partial V_L}{\partial R} = -E_g \frac{\dfrac{\omega^2 M_1 M_2}{R_0}}{\left(R + jX + \dfrac{\omega^2 M_1^2}{R_0} + \dfrac{\omega^2 M_2^2}{R_0}\right)^2} \tag{3.2}$$

and

$$\frac{\partial V_L}{\partial X} = -jE_g \frac{\dfrac{\omega^2 M_1 M_2}{R_0}}{\left(R + jX + \dfrac{\omega^2 M_1^2}{R_0} + \dfrac{\omega^2 M_2^2}{R_0}\right)^2} \tag{3.3}$$

Both expressions are clearly at a maximum when $X = 0$, and thus the cavity should be at resonance for maximum sensitivity. Optimization with respect to the mutual inductances, M_1 and M_2, gives in both cases

$$\frac{\omega^2 M_1^2}{R_0} = \frac{\omega^2 M_2^2}{R_0} = \frac{R}{2} \tag{3.4}$$

By substituting the above three conditions into the expressions for $\partial V_L/\partial R$ and $\partial V_L/\partial X$, the optimum sensitivities, $(\partial V_L/\partial R)_{max}$ and $(\partial V_L/\partial X)_{max}$, that can be obtained with the system are

$$\left(\frac{\partial V_L}{\partial R}\right)_{max} = \frac{-E_g}{8R} \tag{3.5}$$

$$\left(\frac{\partial V_L}{\partial X}\right)_{max} = \frac{-jE_g}{8R} \tag{3.6}$$

Now, the significance of the sensitivities expressed by equations (3.2), (3.3), (3.5), and (3.6) is that when magnetic resonance occurs and the small resistive and reactive impedances δR and δX are coupled into the resonator circuit, the changes in detector voltage are given by $(\partial V_L/\partial R)\delta R$ and $(\partial V_L/\partial X)\delta X$. Moreover, these voltage changes are expressed in a complex form, and the extent to which they effect the modulus of the detector voltage $|V_L|$, to which the detector in fact responds, depends on the phase of the change in voltage δV_L, thus expressed, relative to V_L. The effect of the phase relationship of two a.c. signals of widely differing amplitudes on the modulus of the sum is fully discussed in §2.6. From this discussion it will be clear that, in general, the change in the modulus of V_L is given by $(\partial V_L/\partial R)\delta R \cos\theta_R$ and $(\partial V_L/\partial X)\delta X \cos\theta_X$, where θ_R and θ_X are the phase differences between $(\partial V_L/\partial R)\delta R$ and V_L, and between $(\partial V_L/\partial X)\delta X$ and V_L. Because δR and δX are real numbers, θ_R and θ_X are, respectively, the differences in phase between $\partial V_L/\partial R$ and V_L, and between $\partial V_L/\partial X$ and V_L [see equations (3.1), (3.2) and (3.3)]. This procedure is somewhat complicated,

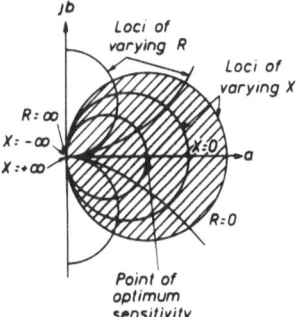

Fig. 3.1. Chart showing dependence of output voltage V_L on the cavity parameters R and X for a transmission-cavity spectrometer. The shaded zone represents $0 < R < \infty$.

and a simpler method enabling the relative degrees of suppression of the absorptive and dispersive components of the magnetic resonance to be determined at a glance, is obtained by plotting the value of V_L on the complex plane, as in Fig. 3.1. To obtain this diagram, express V_L of equation (3.1) as having a real and an imaginary part, $a + jb$. The parameters X and R are then eliminated between the two expressions in turn to obtain equations of a against b, which, when plotted on the diagram, give the loci of V_L as R is varied at fixed values of X, and as X is varied at fixed values of R. In the diagram positive values of R, which are the only values of significance, lie within the shaded area. Then, at any coupling value determined by the ratio of R to $\omega^2 M^2/R_0$, and at any degree of off-tuning, the modulus of δV_L is determined by equations (3.2) and (3.3), and the direction, or phase, relative to V_L is given by the appropriate locus on the diagram. Furthermore, equations (3.2) and (3.3) indicate that the moduli of the sensitivities to absorption and dispersion, $|\partial V_L/\partial R|$ and $|\partial V_L/\partial X|$ respectively, are always equal, and that the phase of the two sensitivities always differs by $90°$. It is therefore possible to transfer the locus described by $\delta R + j\delta X$ in the $(\delta R + j\delta X)$ plane (see Fig. 2.15), since this locus passes through the magnetic resonance directly on to the V_L diagram. Thus, when the magnetic resonance occurs, the tip of the V_L vector traces out, as shown in Fig. 3.2, the circle of Fig. 2.15 superimposed on the diagram of Fig. 3.1, at the value

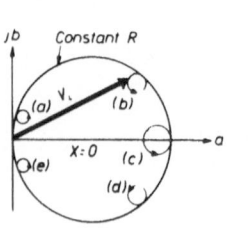

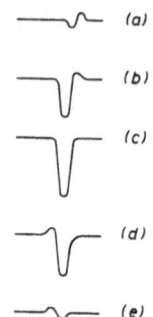

Fig.3.2. Diagram showing the locus of the detector voltage V_L of a transmission-cavity spectrometer as the magnetic resonance is traversed, for various degrees of mis-tuning of the cavity.

Fig.3.3. Variation of detected output from a transmission-cavity spectrometer with the cavity far from resonance on either side (a and e), with the cavity slightly off resonance (b and d), and with the cavity at resonance (c).

of V_L prior to magnetic resonance, and on a scale such that the diameter is equal to the value of $|\partial V_L/\partial R|\delta R$ at the centre of magnetic resonance.

The effect of various degrees of cavity tuning are indicated in Fig. 3.2 (positions $a-e$) and the corresponding variations in $|V_L|$ are given in Fig. 3.3. Thus it is again demonstrated that a transmission cavity at resonance (c) suppresses the dispersive component of the absorption, because small variations in X result only in a shift in phase of V_L, whilst changes in R appear as a change in the modulus of V_L, with no suppression. Furthermore, it is seen that when the cavity is far from resonance, the sensitivity of the system, although much reduced, is largely due to the dispersive component of the magnetic resonance, and the absorptive component is suppressed.

Similar treatments to the above are now applied to the reflection- and absorption-cavity systems, but it is first worth noting, as a practical point, that the optimum coupling factors given by equation (3.4), when inserted into equation (3.1), give a value of V_L at resonance of $E_g/4$, which is exactly half the value that would be obtained if the detector were connected directly to the klystron. Thus, to obtain the optimum coupling factors, the coupling holes should be enlarged until the insertion loss of the cavity is a factor of 2 in voltage, i.e. 6 dB.

§3.2

REFLECTION-CAVITY SPECTROMETER

For the reflection-cavity spectrometer of Fig. 2.12, the detector voltage V_L obtained from the equivalent circuit of Fig. 2.11(e) is $\rho E_g/2$, where ρ is the reflection coefficient of the cavity. From the standard transmission line theory[3] the reflection coefficient is given by

$$\rho = \frac{Z_L - R_0}{Z_L + R_0} \tag{3.7}$$

where Z_L is the impedance presented by the cavity to the line and R_0 the impedance of the line. Thus, as $Z_L = \omega^2 M^2/(R + jX)$

$$V_L = E_g \frac{\dfrac{\omega^2 M^2}{R_0} - R - jX}{2\left(\dfrac{\omega^2 M^2}{R_0} + R + jX\right)} \tag{3.8}$$

The sensitivity to absorption is then

$$\frac{\partial V_L}{\partial R} = -E_g \frac{\dfrac{\omega^2 M^2}{R_0}}{\left(\dfrac{\omega^2 M^2}{R_0} + R + jX\right)^2} \tag{3.9}$$

and to dispersion

$$\frac{\partial V_L}{\partial X} = -jE_g \frac{\dfrac{\omega^2 M^2}{R_0}}{\left(\dfrac{\omega^2 M^2}{R_0} + R + jX\right)^2} \tag{3.10}$$

Then, optimization with respect to the coupling, M, and cavity tuning, X, gives again that $X = 0$ and

$$\frac{\omega^2 M^2}{R_0} = R \tag{3.11}$$

Inserting these conditions into equations (3.9) and (3.10) gives

$$\left(\frac{\partial V_L}{\partial R}\right)_{\text{max}} = \frac{-E_g}{4R} \qquad (3.12)$$

and

$$\left(\frac{\partial V_L}{\partial X}\right)_{\text{max}} = \frac{-jE_g}{4R} \qquad (3.13)$$

By comparing these values with those for the transmission cavity [equations (3.5) and (3.6)], it will be clear that the optimum performance of the transmission cavity is 6 dB below that of the reflection-cavity-circulator system. The effect of the absorptive and dispersive changes in V_L on $|V_L|$ can again be determined from the complex plotting of V_L, which is shown in Fig. 3.4. Now, insertion of the optimum coupling into the expression for V_L gives the result $V_L = 0$. This is in agreement with the general conclusions of §2.5, where it was suggested that the cavity should be matched to the waveguide to obtain optimum sensitivity. However, since $V_L = 0$, the question of the phase relationship of δV_L to V_L does not arise. Thus, neither absorption nor dis-

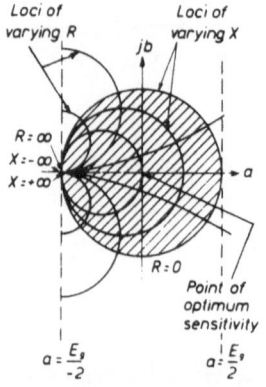

Fig. 3.4. Chart showing dependence of output voltage V_L on the cavity parameters R and X for a reflection-cavity spectrometer. The shaded zone represents $0 < R < \infty$.

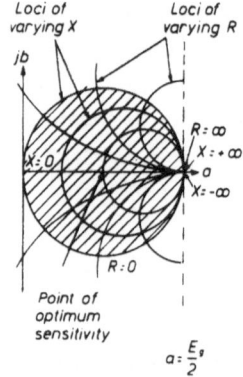

Fig. 3.5. Chart showing dependence of output voltage V_L on the cavity parameters R and X for an absorption-cavity spectrometer. The shaded zone represents $0 < R < \infty$.

persion is suppressed and the signal observed is a mixture of the two (see §2.6). The methods of selecting the absorptive and dispersive components separately are also discussed in §2.6, but a point that is demonstrated by the diagram of Fig. 3.4 is that, even without bucking, dispersion may be suppressed by a small degree of mis-matching of the cavity, whilst absorption suppressed and dispersion observed by matching correctly but slightly mis-tuning. Mis-matching produces a V_L vector which is orthogonal to variations in X and in line with variations in R, whilst mis-tuning produces a value of V_L which is orthogonal to changes in R but in line with changes in X.

§3.3

ABSORPTION-CAVITY SPECTROMETER

An alternative to the transmission-cavity spectrometer is the absorption-cavity spectrometer, of which Fig.2.11(d) is an equivalent circuit. Here the cavity is simply coupled by a single hole to the wall of the waveguide which connects the generator to the detector. When the cavity resonates, energy is drawn from the waveguide, and a reduction in the detector output is observed. It can be seen, from Fig. 2.11(d), that the reduction is due to a reflected impedance of $\omega^2 M^2/$(cavity impedance) being in series with the generator and detector. Thus the detector voltage V_L is given by

$$V_L = E_g \frac{R_0}{2R_0 + \dfrac{\omega^2 M^2}{R + jX}} \tag{3.14}$$

The optimum sensitivity is obtained as before and occurs when $X = 0$ and

$$\frac{\omega^2 M^2}{R_0} = 2R \tag{3.15}$$

which for absorption gives

$$\frac{\partial V_L}{\partial R} = \frac{E_g}{8R} \tag{3.16}$$

and for dispersion

$$\frac{\partial V_L}{\partial X} = \frac{jE_g}{8R} \tag{3.17}$$

The absorption cavity, therefore, has the same sensitivity as the transmission cavity, but with a 180° phase reversal.

Using equation (3.14), we can plot the R and X loci, as in Fig. 3.5. The loci show that, at resonance, dispersion is suppressed. Unlike the transmission cavity, however, dispersion far from resonance is again suppressed, whilst, from Fig. 3.6, it is clear that there is a setting, (b) and (d), on either side of resonance, for which the V_L vector is exactly in phase with variations of X and therefore suppresses absorption and responds to dispersion.

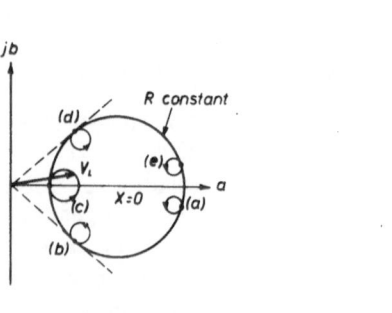

Fig. 3.6. Diagram showing the effect of magnetic resonance on the cavity of an absorption-cavity spectrometer, for various degrees of cavity tuning.

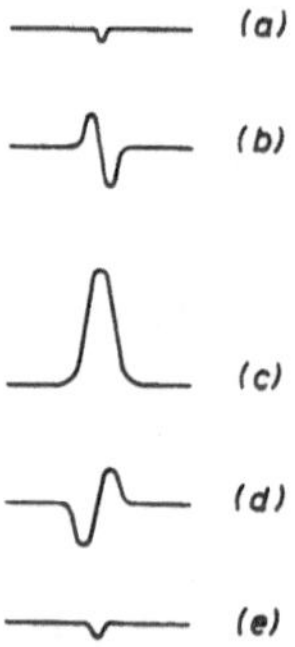

Fig. 3.7. Variation of detected output from an absorption-cavity spectrometer with the cavity far from resonance on either side (a and e), with the cavity slightly off resonance (b and d), and with the cavity at resonance (c). The variation is due to magnetic resonance

Observed spectra for the five positions are shown in Fig. 3.7.

Adjustment of the coupling for optimum sensitivity is obtained for the absorption cavity by increasing the coupling hole until the effect of tuning the cavity to resonance gives a dip of one half in the detector voltage, i.e. an insertion loss of 6 dB. This figure is obtained by inserting the condition of equation (3.15) into equation (3.14), where the result of $E_g/4$ is exactly one half the value of V_L for $X = \infty$.

§3.4

MICROWAVE NOTATION

The practice of using circuit models to represent microwave

circuitry is very common because of the degree to which it simplifies analysis. However, the results so obtained are expressed in terms of voltages, currents and lumped circuit constants, none of which actually exist. Thus, to have significance, the results must be reinterpreted in terms of the corresponding microwave quantities.

The parameters used so far have been the open circuit generator voltage E_g, the waveguide impedance R_0, the detector voltage V_L, the cavity elements R, L and C, the frequency ω and the mutual inductance M representing the coupling to the cavity. Of these, the cavity inductance L is related to the cavity capacitance C, via the resonant frequency of the resonator ω_0, because $\omega_0 = (LC)^{-\frac{1}{2}}$. Thus the cavity reactance X, which is equal to $\omega L - (1/\omega C)$, becomes $\omega_0 L[(\omega/\omega_0) - (\omega_0/\omega)]$. Now $\omega_0 L (= 1/\omega_0 C)$ is related to R by the quality factor Q of the resonator. The Q-factor of a resonator is a fundamental parameter, because, if its value and the value of ω_0 are known, the entire behaviour of the resonator can be determined. Generally speaking, when a resonator is set into oscillation and the driving source is removed, the oscillation decays exponentially because of the losses within the system. Furthermore, it is well known[3] that the time constant T_D of the exponential is equal to twice the energy stored in the resonator divided by the power dissipated in the losses. Thus, if the time constant and the resonant frequency are known, the behaviour of the resonator is completely determined. In fact, the Q-factor is defined as the ratio of (1) the resonant frequency ω_0, to (2) the ratio between power dissipated and the energy stored, so that Q becomes dimensionless. Thus

$$Q = \omega_0 \frac{\text{Energy stored}}{\text{Power dissipated}} = \frac{\omega_0 T_D}{2} \qquad (3.18)$$

Now, in an electrical resonator, the energy oscillates from the inductance to the capacity and is completely stored in the inductance when the current i is a maximum, $\hat{i}$. Thus, the stored energy is equal to $\frac{1}{2}L i^2$, whilst the power loss is equal to $i^2 R/2$. Therefore the Q-factor equals $\omega_0 L/R$, and the cavity reactance becomes $QR[(\omega/\omega_0) - (\omega_0/\omega)]$. When a resonator is loaded, the effect of the load is to reduce the Q-factor of the resonator, which is directly reflected as a reduction in the damping time constant. It is therefore reasonable to define the coupling factor β for each load, to which the resonator is subjected, as the ratio of the power dissipated in the load to that dissipated in the resonator when the resonator is oscillating freely. The meaning of 'oscillating freely' here is not that the driving generator, or any load, is uncoupled, but that the driving force of the generator is reduced to zero. Thus the generator itself imposes some form of load upon the resonator.

Now, in an electrical resonator, the ratio between the power delivered to an external load and the power dissipated within the resonator is equal to the ratio between the resistance reflected by the load into the resonator and the resistance of the resonator itself. Generally speaking, this reflected resistance is equal to $\omega^2 M^2$ divided by the external resistance. Thus, for the transmission cavity, the generator and load coupling factors, β_1 and β_2, are given by $\omega^2 M_1^2/R_0 R$ and $\omega^2 M_2^2/R_0 R$, and both are shown to have an optimum value of $\frac{1}{2}$. For the reflection cavity, β is equal to $\omega^2 M^2/R_0 R$ and has an optimum value of unity, whilst, for the absorption cavity, β is equal to $\omega^2 M^2/2R_0 R$ and again has an optimum value of unity. Alternatively, it is common to express the degree of coupling in terms of an 'external' Q-factor, Q_E, distinguishing at the same time between the loaded and unloaded Q-factors, Q_L and Q_U. Here Q_E is defined by

$$Q_E = \frac{\omega_0 \times \text{Energy stored in resonator}}{\text{Power dissipated in load}} \tag{3.19}$$

Q_U by

$$Q_U = \frac{\omega_0 \times \text{Energy stored in resonator}}{\text{Power dissipated in resonator}} \tag{3.20}$$

and Q_L by

$$Q_L = \frac{\omega_0 \times \text{Energy stored in resonator}}{\text{Total power dissipated}} \tag{3.21}$$

It will thus be clear that

$$Q_L = \left(\frac{1}{Q_U} + \frac{1}{Q_E}\right)^{-1} \tag{3.22}$$

and that

$$\beta = \frac{Q_U}{Q_E} \tag{3.23}$$

When these substitutions are used in any of the foregoing results, the remaining circuit parameter R is found invariably to cancel, and one is left with an expression containing only the generalized parameters. For example, equation (3.1) becomes either

$$V_L = E_g \frac{(\beta_1 \beta_2)^{\frac{1}{2}}}{1 + \beta_1 + \beta_2} \left[1 + j \left(\frac{Q_U}{1 + \beta_1 + \beta_2} \right) \left(\frac{\omega}{\omega_0} - \frac{\omega_0}{\omega} \right) \right]^{-1}$$

(3.24)

or, in terms of Q-factors,

$$V_L = E_g \frac{Q_L}{(Q_{E_1} Q_{E_2})^{\frac{1}{2}}} \left[1 + j Q_L \left(\frac{\omega}{\omega_0} - \frac{\omega_0}{\omega} \right) \right]^{-1}$$

(3.25)

It can still be argued that V_L and E_g are not real voltages, but when it is recalled that $E_g/2$ is the value of the wave propagated towards the cavity system from the generator, the ratio of V_L to $E_g/2$ is simply understood as being the ratio between the field strength of the wave reaching the detector and that emitted by the generator. Thus the response of any of the circuits can now be completely expressed in terms of measurable microwave parameters. The technique of determining the coupling factor by measuring the 'ringing time' of the cavity with and without loading is somewhat impractical however, and it is more common to estimate the degree of coupling from the bandwidth of the cavity resonance. For the particular example of equation (3.25), the effect of varying the frequency ω of the generator through resonance is to make V_L describe the circular plot shown in Fig. 3.1. It is then easily shown that the values of ω for which the tip of the V_L vector is mid-way between the positions for $\omega = \omega_0$ and $\omega = \infty$ (or zero) is that for which $(\omega/\omega_0) - (\omega_0/\omega)$ is equal to $1/Q_L$. Thus, in this case, Q_L may be determined by measuring the frequencies at which the amplitude of V_L falls to $1/\sqrt{2}$ of its value at resonance. It is also worth noting that, for high Q-factors, where ω does not vary greatly from ω_0

$$\frac{\omega}{\omega_0} - \frac{\omega_0}{\omega} \simeq \frac{2\delta\omega}{\omega_0}$$

where $\delta\omega = \omega - \omega_0$. Thus

$$Q_L = \frac{\omega_0}{\Delta\omega}$$

where $\Delta\omega$ is the frequency separation between the 'half-power points' in question. Therefore the degree of coupling can be determined by comparing the loaded bandwidth with the bandwidth at a very light loading. This ratio is then equal to Q_U/Q_L, from which β or Q_E may be determined.

For the other systems, matters are not quite so straightforward, because the circles generated by V_L as the frequency is varied, do not appear at the same point in the diagram. However, it is always true that the separation between the two halfway points on the circle is given by $(\omega/\omega_0) - (\omega_0/\omega) = 1/Q_L$, and, as the form of the diagram is known, Q_L can be determined by modifying the method of interpreting the bandwidth of the resonance line. For example, Fig. 3.4 shows that if a reflection cavity is coupled to present a match when at resonance, the appropriate frequency separation is given by the two klystron frequencies for which the detector voltage is $1/\sqrt{2}$ times the value of the fully reflected wave reaching the detector when the klystron is tuned far from resonance.

§3.5

TRAVELLING-WAVE SPECTROMETER

In §1.4 it was suggested that the cavity resonator used for E.S.R. spectroscopy is a necessary modification of the long waveguide absorption cell used for gaseous microwave spectroscopy. It will therefore be of interest to show that the use of a cavity does not significantly reduce the overall sensitivity below that which would be obtained if an absorption cell could be used for E.S.R. For the absorption cell, V_L is given by

$$V_L = \frac{E_g}{2} \cdot \exp\left(-\gamma_p x_c\right) \tag{3.26}$$

where γ_p is the complex propagation constant $\alpha_p + j\beta_p$, and x_c the length of the cell. If one now reverts to the transmission line equivalent of the system, α_p for a low loss line is given[3] by

$$\alpha_p = \frac{r}{2}\left(\frac{c}{l}\right)^{\frac{1}{2}} \tag{3.27}$$

where r, l, and c are, respectively, the resistance, inductance and capacitance per unit length of the line. Thus

$$|V_L| = \frac{E_g}{2} \exp\left[-\frac{r}{2}\left(\frac{c}{l}\right)^{\frac{1}{2}} x_c\right] \tag{3.28}$$

The effect of the absorption is to add the resistance of the line so that the sensitivity is given by

$$\frac{\partial |V_L|}{\partial r} = \frac{E_g}{2}\left[\frac{-x_c}{2}\left(\frac{c}{l}\right)^{\frac{1}{2}}\right] \exp\left[-\frac{r}{2}\left(\frac{c}{l}\right)^{\frac{1}{2}} x_c\right] \tag{3.29}$$

which is a maximum when $x_c = 2/r \cdot (c/l)^{-\frac{1}{2}}$, giving an optimum sensitivity of

$$\left(\frac{\partial |V_L|}{\partial r}\right)_{max} = -\frac{E_g}{2r} \cdot \exp(-1) \tag{3.30}$$

If the cavity is made from the same waveguide as the absorption cell, it is fair to say that the ratio between the normal resistance per unit length r of the line and the effective resistance per unit length δr reflected into the line owing to the magnetic resonance, is the same as the ratio between the equivalent impedances R and δR in the cavity spectrometer.

Thus a comparison of equation (3.30) with equation (3.12) indicates that the travelling-wave structure gives a sensitivity which, when correctly applied, is $2/e$ times that of the reflection-cavity spectrometer. Thus it is clear that there can, in fact, be a slight gain in changing from a travelling wave to a resonant sample holder.

§3.6

SETTING UP THE REFLECTION-CAVITY SPECTROMETER

In order to consolidate the foregoing remarks it will be useful to describe the normal method of setting up the reflection-cavity spectrometer of Fig. 2.16. Initially, it is required to tune the klystron to resonance, to match the cavity, and to ensure that the klystron, when at resonance, is set to the top of its reflector voltage-output mode. First, the attenuator in the bridge arm is fully inserted to give no drive power except from the cavity. The voltage sweep is then applied to the klystron and the output from the detector coupled to the oscilloscope. If the cavity is nowhere near resonance, the klystron power is simply reflected from the cavity iris to the detector. Consequently the normal klystron mode pattern is observed, as in Fig. 3.8(a). If the sample cavity, or more probably the klystron cavity, is then adjusted to bring the sample cavity resonance within the range of the frequency sweep of the klystron, then the sample cavity will absorb some of the energy and give a dip in the klystron output mode, as in Fig. 3.8(b). The klystron cavity is then adjusted to bring the dip into the centre of the klystron mode. Then, in all probability, the dip observed will not be to zero. This may be because the sample cavity is under- or over-coupled and is best understood by reference to the loci of R and X for the reflection cavity (see Fig. 3.4). In general, as the frequency is swept through resonance, the reflection coefficient is described by a vector joining the origin and one of the circles of constant R. All of these circles start at the point where $a = -E_g/2$ and $b = 0$ for $\omega = 0$ and finish there for $\omega = \infty$. The circle corresponding to exact matching passes through the origin when $\omega = \omega_0$, the resonant frequency. Consequently, the amplitude of the reflection coefficient swings from unity, through zero, and back to unity. Under-coupling corresponds to a circle of smaller diameter than the matched circle, whilst over-coupling corresponds to one of larger diameter. In both cases the modulus never passes through zero. It should, however, be clear that for under-coupling the width of the observed line is somewhat narrower than when matched, whilst for over-coupling the width can be much greater. Indeed, for infinite coupling (no iris) the amplitude of the reflection coefficient remains at unity throughout. Incidentally, if the normal width of the resonance is uncertain, a good method to see whether the cavity is over- or under-coupled, is to introduce slowly a piece of damp fibre into the cavity. If the cavity is under-coupled it will become more so, and the depth of the dip will be reduced, but if the cavity is over-coupled the additional losses will make it pass through the matched condition. Consequently, when the cavity is grossly damped, the dip will first drop down to the base line and then rise to unity. The coupling to the cavity is then adjusted by whatever means is provided until the dip reaches the base line. The cavity is then matched when at resonance. Sufficient power is applied via the bridge arm to provide the optimum drive power to the crystal. The effect of the additional power upon the oscilloscope display depends on the phase relationship of the drive power. If, as is

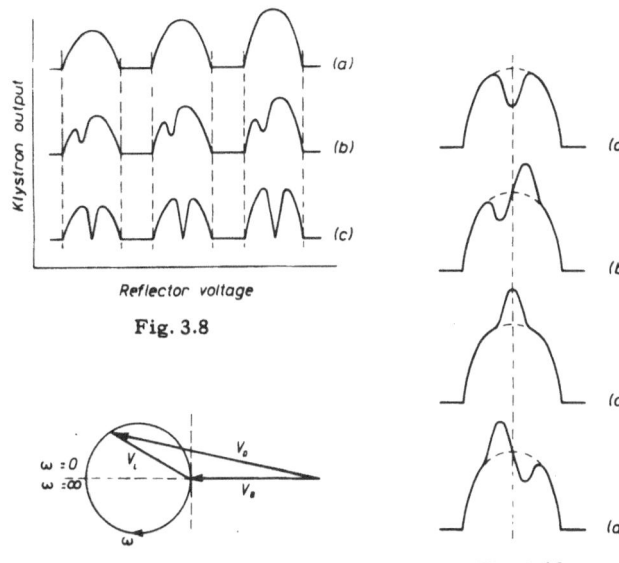

Fig. 3.8

Fig. 3.9   Fig. 3.10

Fig. 3.8. Variation of detector voltage with klystron reflector voltage for a reflection-cavity spectrometer with no bucking; (a) with the cavity off resonance, (b) with the cavity resonance within the range of the klystron frequency sweep, and (c) with the cavity resonance at the centre of the klystron frequency sweep, and with the cavity matched.

Fig. 3.9. Diagram showing the effect of varying the klystron frequency on the detector voltage of a reflection-cavity spectrometer with bucking, with the cavity matched and the bucking phase adjusted for observing absorption. V_B is the bucking voltage, V_L the voltage reaching the detector from the cavity and V_D the combined detector voltage.

Fig. 3.10. Diagrams showing the variation in the output from the microwave detector as the reflector voltage of the klystron is varied; (a) corresponds to the phase relationship of Fig. 3.9 and (b), (c) and (d) to Fig. 3.9 with the bucking voltage phased $90°$, $180°$, and $270°$ ahead of the position of Fig. 3.9.

usual, absorption is to be viewed, the phase of the drive power should align the absorption signal with the real axis of Fig. 3.4, as shown in Fig. 3.9. It should be clear from Fig. 3.9 that such a phase relationship simply lifts the dip of the matched cavity from the base line without altering the position of the minimum.

The effect of other phase relationships may similarly be easily deduced from Fig. 3.9; Fig. 3.10 provides some examples.

The problem of obtaining the exact phase relationship between the bucking and cavity signals for dispersion to be rejected can fortunately be overcome by setting the klystron frequency at minimum (or maximum) in the detected output from the bridge. Then, in any condition of phase, the system is independent, to the first order, of changes in the klystron frequency and therefore of any changes in the reactance of the cavity. Since dispersion is a reactive effect, dispersion is therefore suppressed. Thus, if V_B in Fig. 3.9 is slightly mis-phased, say in the positive direction, a small reduction in the klystron frequency from resonance will bring V_D orthogonal to the V_L circle again and suitably suppress the dispersion. Of course the frequency at which V_D is orthogonal to the V_L circle is also the frequency for minimum V_D. Thus the method of setting up is first to adjust the phase relationship so that the conditions of Figs. 3.9 and 3.10(a) are roughly obtained, and then the sweep is slowly removed and the centre of the cavity resonance maintained in the centre of the display.

If, with this technique, the cavity resonance is not set exactly at the maximum of the klystron mode, the minimum of detector voltage corresponds to the point at which the slope of the microwave bridge and cavity system exactly balances the slope of the klystron mode. Thus it is fairly important to ensure that the initial adjustment of the klystron resonator is made carefully.

§3.7

SPECTROMETER SYSTEMS FOR SATURABLE SAMPLES

So far we have assumed, following the Bloch equations of §2.4, that the susceptibility χ is a constant, i.e. the rotating magnetization M_+ is proportional to the applied rotating field H_+, and thence that the linear component M of M_+ is proportional to the linearly polarized applied field H, of which H_+ is a component. A more exact solution of the Bloch equations shows that this is not so, and that the proportionality only holds for values of applied field below a certain critical (or 'saturation') value. Before quoting the Bloch result we shall consider the simple two-energy-level model of the system to obtain an idea of what to expect. Normally, for any two-level system, the lower level is, in accordance with the Boltzmann distribution of equation (1.6), the more densely populated level. Consequently, when radiation of the appropriate frequency is applied, more upward transitions than downward ones are stimulated. This tends to increase the population of the upper level and, if allowed to continue, eventually results in the populations of the two levels being equal. No further increase in the upper populations then ensues, because equal populations mean an equal number of upward and downward transitions. In this situation no further power is absorbed from the applied field. However, the longitudinal relaxation mechanisms tend to depopulate the upper level and restore the Boltzmann distribution. If the effect of the strength of the applied radiation is small compared with that of the relaxation mechanisms, then normal absorption is observed. If, on the other hand, the applied radiation is strong enough to make the levels nearly equal, then the only power absorbed is that necessary to restore the power removed by the relaxation mechanisms. This power is constant and so the power absorbed from the applied field no longer increases with the intensity of the field.

Reverting to the Bloch model of §2.4, where it was shown that the power absorbed is proportional to the product of M_+ and H_+, it now becomes clear that, for full saturation, M_+ must vary inversely with, rather than in proportion to, H_+. From equation (2.14) the exact Bloch equation for χ'' at resonance is

$$M = H\chi'' = \frac{-\pi H T_2 g \beta M_z}{h} \qquad (3.31)$$

where M_z is the static component of magnetization in the direction of the applied d.c. field and is assumed to be equal to the value M_0 when there is no applied rotating field. In fact

$$M_z = M_0 \left[1 + T_1 T_2 \left(\frac{\pi g \beta}{h} H \right)^2 \right]^{-1} \qquad (3.32)$$

So by combining equations (3.31) and (3.32) we again find that below saturation magnetization is proportional to applied field, and above saturation it is inversely proportional to it. The significance of this effect is to modify the conclusions of §2.5 regarding the two requirements of the microwave system of a spectrometer. These are:

   (i) For maximum sensitivity the available power from the klystron should be used to develop the highest possible microwave magnetic field at the sample.

  (ii) The rotating magnetization of the sample should be coupled via the cavity to the detector in such a way as to propagate the greatest amount of power to the detector.

When saturation is possible, it is no longer necessary, or even desirable, to provide the maximum field at the sample. All that is needed is to provide the field corresponding to the maximum magnetization, and provided that this remains possible for the given klystron, the first requirement may be neglected.

When applying the modified criterion to the transmission cavity, it is clear that the primary requirement of coupling the largest amount of power to the detector is satisfied by making the coupling to the generator small and matching the detector to the cavity. Provided that the generator coupling does not become so small that saturation is no longer possible, there is then no difference between the sensitivities of the transmission and reflection systems.

Unfortunately, the other disadvantages encountered with the transmission cavity, such as the inability to adjust the level of microwave drive to the detector or to view dispersion when the cavity is at resonance, still remain. However, the bucking arrangement of Fig. 3.11 solves effectively these problems. Since, from Fig. 2.24, the optimum microwave drive is never more than half a milliwatt, sufficient drive from a 50 mW klystron can be obtained by using two 10 dB couplers, and this incurs negligible loss of the output signal. To obtain optimum microwave drive for observing absorption, the phase relationship of the bridging signal is adjusted to lie along the real axis of Fig. 3.1 and, thus, to add or subtract from the signal from the cavity at resonance, depending on whether the power reaching the detector via the cavity is in phase or in anti-phase with that from the bucking arm.

To observe absorption the setting-up procedure is similar to that for a reflection cavity. When observing dispersion, however, it is slightly more difficult to obtain full sensitivity, for the combined signal from the cavity and the bridging arm is required to produce a vector which is in line with the circle of variable

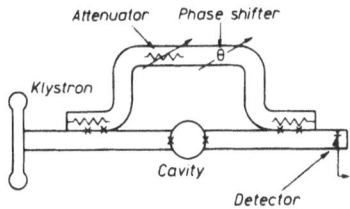

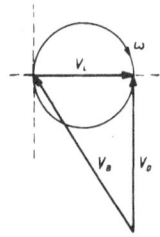

Fig.3.11. Transmission-cavity spectrometer with a bucking arm.

Fig. 3.12. Diagram showing the correct phasing of the bucking signal V_B in the system of Fig. 3.11 to produce an input V_D to the detector which responds to the dispersion signal only. V_L represents the signal from the cavity.

X in Fig. 3.1. The required phase relationship is illustrated by Fig. 3.12. A further simplification, only possible when saturation occurs, is the replacing of the circulator of the reflection-cavity system of Fig. 2.16 with a directional coupler, either as in Fig. 3.13 or as in Fig. 3.14. Of these, Fig. 3.13 is the direct replacement, whilst Fig. 3.14 is a simplification. In both arrangements replacement of the circulator by the directional coupler need give no loss when saturation is possible. Provided the coupling coefficient is not stronger than about 10 dB, the loss of signal power from the detector is negligible, and presumably enough power still reaches the cavity to give the permitted degree of saturation. In Fig. 3.13, the drive power to the crystal is provided in the normal way, whilst, in Fig. 3.14, it is provided by a signal reflected from the fourth arm of the directional coupler. The amplitude and phase of this reflected signal may be controlled to provide optimum drive power and to resolve absorption and dispersion. The control of amplitude and phase of the reflected signal can be achieved in several ways. One is to use a matched load and a slide screw tuner so that the penetration of the screw determines the amplitude of the reflected wave and its position the phase. Slide screw tuners tend to be unsatisfactory unless thoroughly choked. A better method is to use an adjustable short circuit to control the phase of the reflection, and an attenuator to control the amplitude. Alternatively, a fixed short

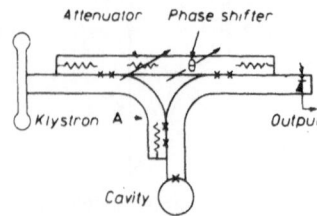

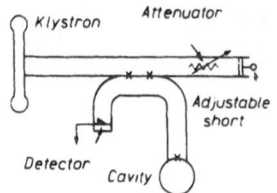

Fig. 3.13. Replacement of the   Fig. 3.14. Simplification of the
circulator in a reflection-cavity   arrangement of Fig. 3.13.
spectrometer by a directional
coupler.

circuit and phase shifter may be used with equal success.

The main weakness of the method of Fig. 3.14 is that it is not possible to provide a crystal drive power greater than the power fed to the cavity. This means that, for very easily saturable samples, the crystal drive power may have to be well below optimum. One solution is to put an adjustable ferrite isolator in the cavity arm. An adjustable isolator gives a variable attenuation for one direction of propagation and no attenuation in the other. Thus, by making the attenuating direction the direction towards the cavity, the power fed to the cavity can be reduced below the detector drive power with no loss in the signal propagated to the detector. Here one is simply substituting one ferrite device for another, so that the difference in cost would probably be very slight and the move seems unjustified. Furthermore, one other reason for not using ferrite devices is that they sometimes tend to be influenced by the field of the main magnet. Thus the arrangement of Fig. 3.13 is probably better, since here, by adding an attenuator at point A, the cavity power can be reduced to any level without limiting the klystron drive to the detector.

§3.8

COMPROMISE SYSTEMS

Where saturation is not possible, it is interesting to calculate what would be the optimum coupling coefficient for the directional coupler used in Figs. 3.13 and 3.14. If the power coupling factor is a_c, then there will be a loss of a_c in feeding the power from the klystron to the cavity and a loss of $(1 - a_c)$ in feeding the signal power from the cavity to the detector. Optimizing the total loss of $a_c(1 - a_c)$ gives $a_c = 1/2$ and a total loss of $1/4$. Thus a 3 dB coupler should be used, or alternatively a magic-tee, to give the arrangement of Fig. 3.15, with possibly an adjustable isolator in the cavity arm. The magic-tee and reflection-cavity system is in fact quite popular. The loss of $1/4$ is a power loss, so that the voltage loss is $1/2$, which makes the system equal in sensitivity to the absorption and transmission systems for unsaturable samples. For a compromise design for a spectrometer which is likely to be used for saturable and unsaturable samples, it would seem reasonable to use a coupling factor which, for both types of sample, degrades the system by equal amounts

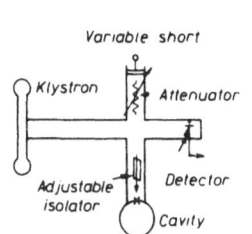

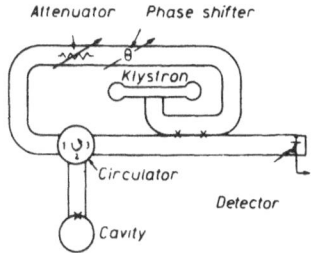

Fig. 3.15. Spectrometer using a magic-tee

Fig. 3.16. Reflection-cavity spectrometer with bucking, for klystrons of reduced power.

from the optimum. For unsaturable samples, the optimum loss is $1/4$, whilst, in general, the loss is $a_c(1 - a_c)$. For saturable samples, the optimum loss is unity, whilst the actual loss is $1 - a_c$, which is the loss due to the directional coupler transmitting power from the cavity to the detector. Then, making the ratio between the actual loss and the optimum loss equal in the two cases gives

$$\frac{a_c(1 - a_c)}{1/4} = \frac{(1 - a_c)}{1}$$

which gives a value of $a_c = 1/4$ or a coupling factor of 6 dB. For both samples the additional power loss over the circulator system is $3/4$, or a voltage loss of $\sqrt{3}/2$, which is a very small sacrifice. Furthermore, in practice, the majority of samples are saturable, and in such cases the above figure is the overall loss relative to a circulator system. For unsaturable samples the power loss over the circulator system is 3/16, i.e. a voltage loss of $\sqrt{3}/4$, which might more strongly justify the circulator.

Since the optimum drive power to the crystal is about 20 dB below the power available from a 50 mW klystron, two 10 dB couplers would seem to be most suitable for the bridge arm of the normal reflection-cavity-circulator system of Fig. 2.16. For lower klystron powers, a more economical arrangement[4] is that of Fig. 3.16, where the directional coupler gives exactly the right crystal drive level. The phase shifter then adjusts the phase relationship between the sample signal and the drive, and the attenuator controls the cavity power.

An even simpler arrangement is to use no bridging at all, as in Fig. 2.12, and to obtain the klystron drive to the crystal by mismatching or off-tuning the cavity. This approach, discussed in §3.2, only works when the required crystal bias power is small compared with the power incident upon the cavity.

It should be noted that, for saturable samples, it is rare to operate at the degree of saturation, which gives the maximum degree of magnetization. Usually the applied field is increased to the point just before the magnetization ceases to be proportional to the applied field. However, the basic consideration of having a fixed maximum magnetization is not altered, the maximum value being just reduced. One reason for avoiding any degree of saturation is that the tendency to saturate is reduced as the field is varied from the centre of the resonance line, so that the centre is attenuated relative to the wings, and the line is broadened. A further error results if the susceptibility of the sample is determined by comparison with a standard material. The two will then, in general, exhibit different degrees of saturation, and thus give an incorrect comparison of intensities.

§3.9

TRANSLATION OF
KLYSTRON FREQUENCY-MODULATED NOISE

In §2.7 we considered how the klystron can be amplitude-modulated by noise, and how this can add to the noise at the output of the detector. The results of Fig. 2.24 show that, for a normal double-channel-mixer detecting system of the type used for double-modulation recording, the contribution of klystron amplitude-modulated (a.m.) noise is generally small. The klystron is also frequency-modulated by noise and this frequency-modulated (f.m.) noise may in some cases contribute to the overall level. If, for example, the transmission-cavity spectrometer of Fig. 1.4 is incorrectly set up, so that the klystron is tuned to one side of the cavity response curve, then variations in klystron frequency are converted to variations in the amplitude of the transmitted signal, using the sloping edge of the cavity response curve as a discriminator. In practice, the conversion of f.m. to a.m. noise in such a case is usually efficient enough to increase the overall noise of the system considerably. It is therefore important not only in order to avoid mixing dispersion with the absorption signal, but also to avoid decreasing the signal-to-noise ratio, that the cavity should be tuned accurately to resonance. In fact, the method of setting up described in §1.5 is one which uses the criterion of minimum sensitivity of the system to klystron-frequency modulation to indicate that the top of the cavity response has been reached. Thus, provided this method is followed, no klystron f.m. noise should be observed.

For any of the spectrometer systems so far discussed, the condition for insensitivity to dispersion is the condition for insensitivity to changes in cavity reactance. This is also the condition for insensitivity to changes in frequency, and, provided the cavity is the only frequency-sensitive element in the system, the condition for observing pure absorption always coincides with the condition for full suppression of klystron f.m. noise.

Conversely, the condition for maximum sensitivity to dispersion is also that of maximum sensitivity to changes in klystron frequency, so that any of the above systems, when set to observe dispersion, will also be very prone to klystron f.m. noise. Methods of overcoming this problem for dispersion spectrometers are discussed in §§3.12 and 3.13.

For absorption spectrometers it is so important to ensure that the cavity system is tuned exactly, and that some kind of automatic frequency control (a.f.c.) system is used. One or two such systems are discussed in Chapter 6.

§3.10

ENHANCED AMPLITUDE-MODULATED NOISE

In a reflection-cavity system it is often possible for the cavity to exaggerate the effects of klystron a.m. noise. The phasor model of a noise-modulated klystron is best used to illustrate this point. If, as in Fig. 3.17(a), the klystron carrier is represented by a phasor rotating at the angular frequency of the klystron ω, then amplitude modulation by a noise component of frequency ω_n will result in sidebands of frequency $\omega + \omega_n$ and $\omega - \omega_n$ being placed on either side of the carrier. When these are viewed from the carrier vector they appear as additional phasors, which rotate in opposite directions relative to the

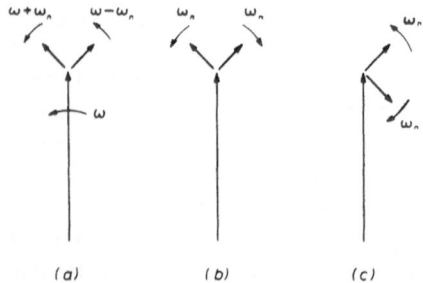

Fig. 3.17. Phasor representation of a modulated carrier (a) for amplitude modulation at a frequency ω_n, (b) as for (a) but viewed from the carrier phasor, and (c) for frequency modulation viewed from the carrier phasor.

carrier phasor at the modulating frequency ω_n. Furthermore, they must be phased so that their resultant lies in the direction of the carrier phasor and is able, periodically, to reduce and increase the amplitude of the carrier phasor. It is also possible to view f.m. noise in a similar way, because, if the sideband phasors are phased so that their resultant is always orthogonal to the carrier, the effect will be to modulate the phase, and therefore the velocity, i.e. frequency, of the phasor. Using this picture, the reason why the cavity does not convert f.m. noise to a.m. noise when it is exactly at resonance is that, with say the

transmission cavity, the carrier and sidebands appear on the
R and X plot of Fig. 3.1, as shown in Fig. 3.18(a). Here the
upper sideband is advanced in phase and the lower sideband is
reduced. Also, the amplitudes remain equal. Thus the resultant
of the two sidebands remains orthogonal to the carrier. If the

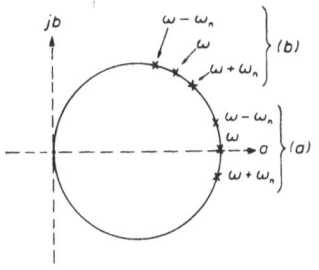
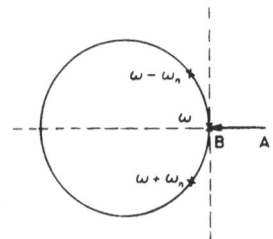

Fig. 3. 18. Diagram showing the
response of a transmission-
cavity to a noise frequency-
modulating component of frequency ω_n (a) with the cavity
at resonance and (b) with the
cavity not at resonance.

Fig.3. 19. Diagram showing the
accentuation of klystron a.m.
noise sidebands by a reflection-cavity spectrometer.

cavity is not at resonance, as in Fig. 3.18(b), the phase relationship of the resultant is not altered much relative to the carrier,
but the amplitudes of the two sidebands become unequal. The
resultant of the sidebands then follows an elliptical path instead
of a linear one, so that the carrier is amplitude-modulated as
well as phase-modulated.

The phasor model may now be applied to the reflection-cavity
spectrometer of Fig. 2.16 to show that, even when the cavity is
tuned to resonance and the bucking signal phased to observe
absorption, the a.m. noise is accentuated. In Fig. 3.19, the circle
represents the reflection from the klystron to the detector by the
cavity, whilst AB represents the bucking signal. Since the reflection at resonance is zero, the carrier term is therefore trans-
ferred to the detector with an efficiency proportional to $|AB|$, but
the sidebands, appearing as shown, are also reflected in part by
the cavity and so reach the detector with an increased efficiency
and a phase shift. From Fig. 3.17 it is clear that the phase shift
has no effect on the depth of modulation, but that the increased

transmission of the sidebands results in a greater depth of noise modulation. This increase can sometimes be of practical significance and is discussed further in §3.14.

§3.11

BALANCED MIXER SPECTROMETER

A useful modification[5] of the reflection-cavity-circulator spectrometer of Fig. 2.16 is to combine the microwave drive and the signals from the cavity in a balanced mixer which uses two crystals as shown in Fig. 3.20. Here the balanced mixer can be of any of the usual types. Either a magic-tee, as shown, or a 3 dB directional coupler can be used, the main point being that

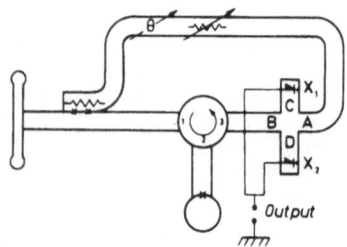

Fig. 3.20. Reflection-cavity spectrometer with a balanced mixer.

the relative phases of the signals entering ports A and B are reversed for the two output ports C and D. Thus, if the signal reaching each crystal from the cavity by port B is the same, the $X - R$ plot for the matched cavity is the same for each (see Fig. 3.21). The drive power entering port A reaches crystals X_1 and X_2 with opposite phase so that, if the phasing is adjusted to observe absorption, the drive signal for one crystal is given by V_{AC} and for the other by V_{AD}. If the crystals are arranged to have opposite polarities and the two outputs are taken in parallel to the amplifier, absorption in the cavity causes a decrease in the output of crystal X_1 and an increase in that of crystal X_2. Consequently, the balance of the two opposed crystal outputs is disturbed and an output signal develops. One advantage of the method is that, in the absence of any signal from the cavity, there is no net output from the two crystals. Hence any amplitude

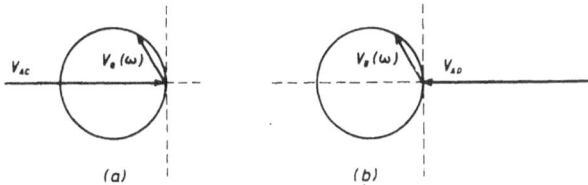

Fig. 3.21. Diagram showing how the bucking signals, V_{AC} V_{AD}, and the signal V_B reflected from the cavity combine in the two detector arms of the balanced mixer of Fig. 3.20.

modulation of the klystron has no effect and klystron a.m. noise is suppressed. The method works equally well if crystals of similar polarity are used, but then a difference amplifier, probably in the form of a balanced transformer, is needed to take off the output. It is well known that the balanced mixer arrangement suffers no loss on account of the different circuit arrangements, and has the same basic noise factor, apart from the reduction of klystron a.m. noise, as a normal single-ended mixer. In fact, the ability to reduce the normal klystron a.m. noise is not the main advantage of the balanced mixer in this particular application. Indeed, Fig. 2.24 suggests that normal klystron a.m. noise does not often contribute significantly to the overall noise level at all. The main advantage of the method is in the reduced sensitivity to klystron f.m. noise. The point is demonstrated by Fig. 3.22 where the cavity is assumed to be slightly off-tune. In a conventional system, frequency modulation of the klystron then causes

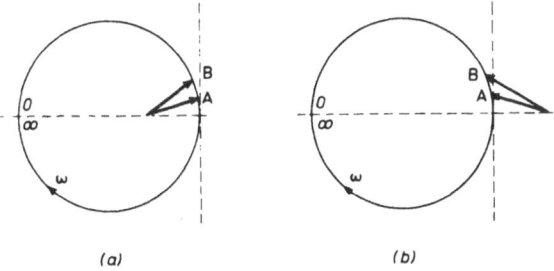

Fig. 3.22. Diagrams showing the modulation of input voltage to the individual detectors in the balanced mixer spectrometer of Fig. 3.20 as the klystron is frequency-modulated, the cavity being off-tune.

the detector signal to oscillate, say from A to B, and because the system is off resonance, some amplitude modulation results. In the balanced mixer system, however, the output from the second crystal is amplitude-modulated to an almost equal extent. Consequently, even when off-tune, the system is relatively insensitive to klystron f.m. noise. The same arguments apply to the ability to suppress dispersion, so that the cavity may be allowed to drift further from resonance before the absorption signal begins to be distorted. It will be clear that the above arguments apply most completely when the drive to the cavity is large compared with that from the drive arm. Fortunately this is the condition for which klystron noise is most troublesome.

Unfortunately, balanced mixer detection does nothing to reduce the effect of enhanced a.m. klystron noise, because this arrives at the mixer via the cavity and not via the drive arm. The effect is illustrated by Fig. 3.23. Here AC and AD represent the bucking signals reaching the detectors, whilst the circles show how the signal reflected from the cavity would vary if the klystron were detuned. The noise sidebands are, of course, off-tune and are therefore reflected with amplitudes and phases proportional to CX_1 and CX_2 for the first detector, and DX_1 and DX_2 for the second. The upper and lower noise sidebands f_{nu} and f_{nl} in (c and d) therefore reflect these quantities and combine with the bucking signals as shown. It is thus clear that the amplitude modulation so produced is in antiphase for the two detectors. Hence the balance between the outputs is disturbed and the signal is transferred to the amplifier.

§3.12

TWIN-CAVITY SPECTROMETER

A good method of avoiding both klystron f.m. noise and enhanced a.m. noise is the twin-cavity system[6] of Fig. 3.24. This is effectively a transmission-cavity spectrometer, but with a second cavity in a balancing arm. The input power, split by a magic-tee, passes through two identical cavities, recombines at the output, and is fed to the detector. With this arrangement, no power at any frequency can reach the detector from the klystron provided that the cavities are identical and are identically tuned

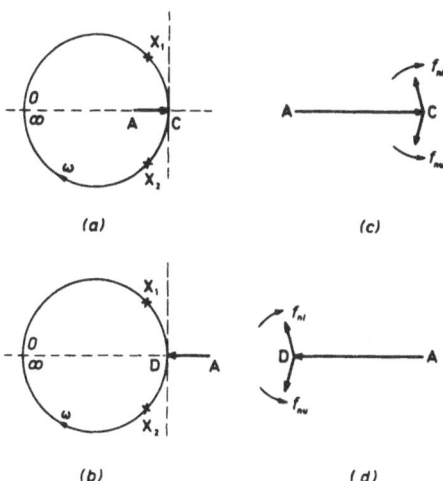

(a)   (c)

(b)   (d)

Fig. 3.23. Diagram showing why enhanced a.m. noise is not removed by the balanced mixer. (a) and (b) Variation with klystron frequency of the signal reaching each detector; X_1 and X_2 represent noise sideband frequencies. (c) and (d) Phasor diagrams showing how the noise sidebands combine with the bucking signals at the two detectors.

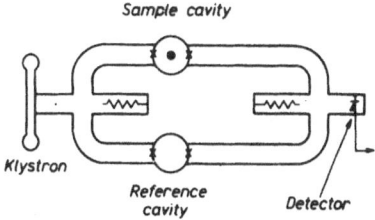

Fig. 3.24. Twin-cavity spectrometer.

and coupled. Thus all types of klystron noise are precluded. One of the cavities is made the sample cavity, so that any absorption will unbalance the bridge and deliver a signal to the detector. It is clear that, with the system indicated, there will be a 3 dB loss at the input and output tees, and so a loss of 6 dB relative to a normal transmission cavity results. It is therefore better to replace the magic-tees by normal matching power dividers. The precise improvement in sensitivity and the optimum couplings

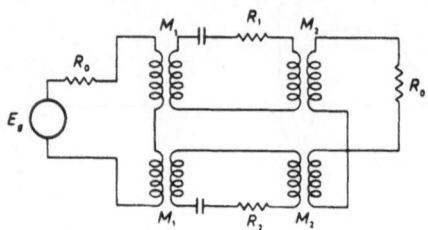

Fig. 3.25. Equivalent circuit for a twin-cavity spectrometer in which power dividers have replaced the magic-tees.

into the cavities are now not obvious, so some analysis must be carried out. The equivalent circuit of the modified arrangement is shown in Fig. 3.25, where all the symbols conform to the notation of §3.1. Clearly, either the input or the output couplings must be in opposition to obtain the required degree of balancing. The particular choice is immaterial and M_2 is chosen. Then the detector voltage V_L is given by

$$V_L = E_g \omega^2 M_1 M_2 (R_2 - R_1)[4\omega^4 M_1^2 M_2^2$$
$$+ R_0 (R_1 + R_2)(\omega^2 M_1^2 + \omega^2 M_2^2) + R_0^2 R_1 R_2]^{-1} \qquad (3.33)$$

Since M_1 and M_2 are interchangeable, it is clear that M_1 equals M_2 in any optimization. We therefore put $M_1 = M_2 = M$ in the expression for the sensitivity of the system $\partial V_L / \partial R_2$, which gives when we subsequently put $R_1 = R_2$

$$\frac{\partial V_L}{\partial R_1} = E_g R_0 \left(4\omega^2 M^2 + 4R_0 R_1 + \frac{R_0^2 R_1^2}{\omega^2 M^2}\right)^{-1} \qquad (3.34)$$

and for optimum coupling

$$\omega^2 M^2 = \frac{R_0 R_1}{2}$$

Thus the optimum sensitivity is given by

$$\frac{\partial V_L}{\partial R_1} = \frac{E_g}{8R_1} \qquad (3.35)$$

This equation is the same as that for the normal transmission-cavity spectrometer, so that the loss of the magic-tees is made good, and furthermore, the optimum cavity couplings are the same. Clearly, to provide the detector drive some kind of bridge

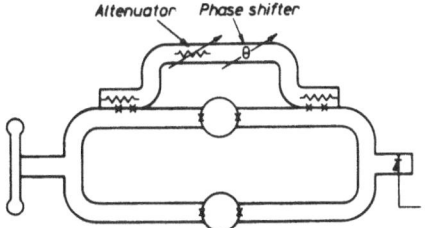

Fig. 3.26. Twin-cavity spectrometer with bucking.

arm is required; the full arangement is shown in Fig. 3.26. When only a small amount of drive to the detector is needed, it is possible to obtain the drive by mis-matching or mis-tuning the reference cavity depending on whether absorption or dispersion is to be observed.

§3.13

INDUCTION-CAVITY SPECTROMETER

A promising alternative to the twin-cavity spectrometer is the induction-cavity or crossed-mode-cavity system.[7] The principle of this approach is derived from N.M.R. spectroscopy, where the induction system of Fig. 3.27 is frequently used. Here, instead of using one coil to apply the radio-frequency magnetic field and to pick up the signal from the rotating magnetization of the sample, two separate coils are used. These are arranged with their axes mutually perpendicular and both are made perpendicular

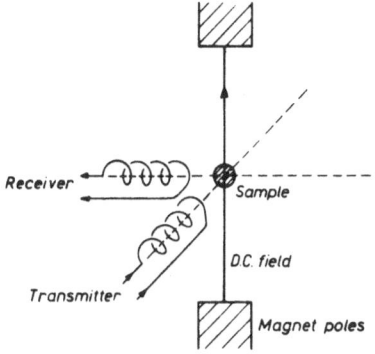

Fig. 3.27. Principle of an induction spectrometer.

to the direction of the d.c. magnetic field. Hence, the transmitting and receiving coils are set in the plane of the rotating magnetization of the sample, and the receiving coil picks up the magnetization induced by the transmitting coil without actually picking up any component directly from the transmitting coil. Thus, at any frequency, the transmitting and receiving coils are only coupled via the sample, and any noise superimposed upon the klystron supply is effectively rejected.

The microwave equivalent of this idea is to use a cavity which has two orthogonal modes occurring at the same frequency. Two arrangements have so far been investigated, the H_{011} rectangular cavity of Fig. 3.28(a) and the H_{111} cylindrical cavity of Fig. 3.28(b). The frequency f_0 at which the rectangular cavity resonates is given by

$$f_0 = c\left[\left(\frac{1}{2l}\right)^2 + \left(\frac{1}{2a}\right)^2\right]^{1/2} \qquad (3.36)$$

where c is the velocity of light and a, and l are the dimensions of the cavity. The main point of the significance here is that f_0 is independent of one dimension, b. Thus if b is made equal to l, as in Fig. 3.28(c), it is possible to make the cavity resonate at the same frequency in two distinct and orthogonal modes. Moreover, if the sample is placed as shown, the magnetic field directions of the two modes at the sample are perpendicular. Thus energy may be coupled from the generator into one mode and from the other into the detector. The arrangement of Fig. 3.28(c) suits the purpose fairly well, the coupling holes to each mode being arranged at positions of zero magnetic field for the unwanted mode and at maximum magnetic field for the wanted mode. It is also generally necessary to have some kind of fine tuning control to make the two modes exactly coincident in frequency. Thus, two dielectric tuning screws are shown and these again must be arranged to be at maximum electric field for the mode that is to be influenced, and at zero electric field for the other mode. The principal problem with such an arrangement is the lack of symmetry of the receiving system relative to the transmitting system. This tends to perturb the normal mode pattern and couples the transmitting mode to the receiving coupling holes unless further tuning screws are added to compensate for

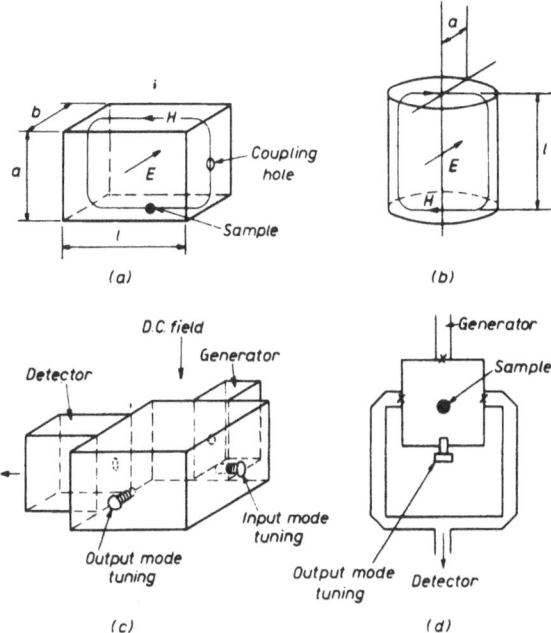

Fig. 3.28. Induction cavity systems. (a) Field pattern in an H_{011} rectangular cavity. (b) Field pattern in an H_{111} cylindrical cavity. (c) Simple induction-cavity spectrometer. (d) Induction cavity of improved symmetry.

the effect. In consequence, symmetrical coupling arrangements, such as that shown in Fig. 3.28(d), are somewhat preferable. With the system shown, only the receiving mode is tunable, and it is assumed that the klystron can be adjusted to bring the transmitting mode to resonance. The cavity of Fig. 3.28(b) is the cylindrical equivalent of Fig. 3.28(a) and can be used in much the same way.

The reason why induction-cavity spectrometers do not enjoy a great deal of popularity is that they tend to be more bulky than simple cavity systems, and it is difficult to fit an induction cavity, particularly of the type employing many tuning screws, into the magnet gap. Also, tuning can be very inconvenient and often it is not possible to use effectively as large a sample as used for normal systems. Neither is the problem of inserting the sample easily solved. However, for potentially low noise systems, such as the maser and parametric amplifier spectrometers of

Chapter 7, the induction cavity may prove essential. Of course, with the induction system, some sort of bucking arm is again required to obtain suitable drive to the detector and to discriminate between the absorption and dispersion signals. In practice, the appropriate overall system is obtained by replacing the cavity and circulator of the reflection-cavity spectrometer of Fig. 2.16 by the induction cavity.

§3.14

CHOICE OF MODULATING FREQUENCY IN A DOUBLE-MODULATION RECORDING SYSTEM

In §2.3 it was stated that 100 kc/s was a common choice for the frequency of d.c. magnetic field modulation for a spectrometer with a double-modulation recording system. Now it is possible to discuss the reasons for this choice.

The first factor to be considered is that of line broadening. It has been shown in §2.3 that the amplitude of the field modulation used for double-modulation recording must be small compared with the linewidth, if broadening and distortion are to be avoided. It is also necessary, however, to limit the modulation frequency f_m to a value which is small compared with the linewidth when expressed as a frequency. The effect is much easier to understand when the klystron frequency is modulated instead of the magnetic field. Suppose, for example, the width $(\Delta f)_E$ of the resonance line is 1 Mc/s, corresponding according to equation (1.5) to a figure of 300 milligauss, at an operating frequency of 10 Gc/s. We are now demanding that, in addition to the frequency deviation f_d of the klystron being small compared with $(\Delta f)_E$, the modulation frequency f_m must also be small. Consider, then, the extreme case where the amplitude requirements are satisifed but the value of f_m is much in excess of $(\Delta f)_E$, i.e. $f_d = (\Delta f)_E/\alpha$, where $\alpha \gg 1$, and f_m is large compared with $(\Delta f)_E$, i.e. $f_m = \beta(\Delta f)_E$, where $\beta \gg 1$. Now, the expression for a frequency-modulated wave of voltage v is

$$v = \hat{v} \cos \phi(t) \qquad (3.37)$$

where $\hat{v}$ is the peak value of v and $\phi(t)$ is a time-dependent phase angle.

For the present case

$$\phi(t) = \int(\omega + \omega_d \cos\omega_m t)dt \qquad (3.38)$$

where ω is the frequency of the klystron when unmodulated, $\omega_d = 2\pi f_d$ and $\omega_m = 2\pi f_m$. Then

$$v = \hat{v}\cos\left[\omega t + \frac{f_d}{f_m}\cos\left(\omega_m t - \frac{\pi}{2}\right)\right] \qquad (3.39)$$

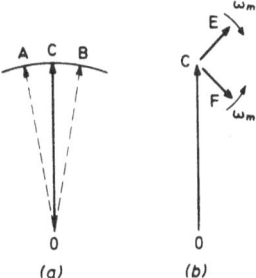

Fig. 3.29. (a) Vector representation of a f.m. wave. (b) Representation of (a) by carrier and sidebands.

Now, f_d/f_m represents the peak deviation in phase and is equal to $(\alpha\beta)^{-1} \ll 1$. Thus the wave can be represented vectorially, as in Fig. 3.29(a), by a phasor rotating at a mean frequency of ω_0 and swinging sinusoidally from position OA to position OB to describe an angle of $2(\alpha\beta)^{-1}$, which is small. AB then approximates to a straight line and the entire system can be synthesized, as in Fig. 3.29(b), by the carrier and two rotating phasors CE and CF of frequencies $\omega + \omega_m$ and $\omega - \omega_m$. In this situation the description of Fig. 2.4 breaks down completely because there one assumes that, when the frequency of the resonator is changed relative to the generator, the level of oscillation falls immediately to its new equilibrium condition. In fact, the time the system takes to reach the equilibrium condition is inversely proportional to the system bandwidth and so the model is only correct for changes in frequency of lower frequency than the bandwidth of the resonator. In the present case the reverse is true, and the normal process of absorption must be understood as follows. First, the f.m. generator can be approximated by a carrier, and two sidebands that are separated from the carrier by the modulation frequency f_m, which is large compared with the resonance linewidth (Δf). Thus, as the frequency of the carrier and sideband system is slowly varied past the frequency of the E.S.R. resonance, each component is absorbed separately. Now as the carrier is absorbed, the actual absorption has no effect, since reducing OC still leaves the wave only frequency-modulated. In fact, it is the dispersion which is effective in producing the amplitude modulation at ω_m which the microwave detector detects. The dispersive voltage appears in quadrature with OC and thus slightly extends CB in Fig. 3.29(a) and slightly reduces AC. Thus the lengths of OA and OB are no longer equal, so the wave is amplitude-modulated. Thus amplitude modulation with a depth proportional to the dispersive signal results. This, of course, is the familiar S-curve normally given by the derivative of the absorption curve, so the observed output from the spectrometer seems to be unaltered. If now the carrier frequency is adjusted so that one or other of the sidebands fall at the absorption frequency, then one of the sidebands is absorbed. This is equivalent to adding a small rotating sideband vector in opposition to, say, CE of Fig. 3.29(b). Thus the linear path ACB becomes elliptical, amplitude modulation results, and instead of one resonance line being observed, three are. These lines themselves will not be broadened, but, for values of modulation frequency comparable with the linewidth, the tendency to split will produce broadening. Moreover, normally one would only observe the two sideband resonances for the extreme case discussed above. Clearly, for sideband resonances, the peak of the amplitude modulation occurs when the modulated carrier is in the central position, i.e. when the carrier is rotating at its fastest relative to the unmodulated position, and therefore at the instant of maximum frequency deviation. Thus, the amplitude modulation is in phase with the frequency modulation for one sideband and in antiphase for the other. Hence, one sideband appears as an absorption peak and the other as an absorption dip on the pen recorder. For the carrier absorption the peak amplitude modulation occurs when the vector is at its extremities OA and OB. This is the position of peak phase deviation and zero frequency deviation. Thus the phase of the amplitude modulation is in quadrature with the frequency modulation and therefore does not give an output from the phase-sensitive detector. If the phase of the reference is shifted by

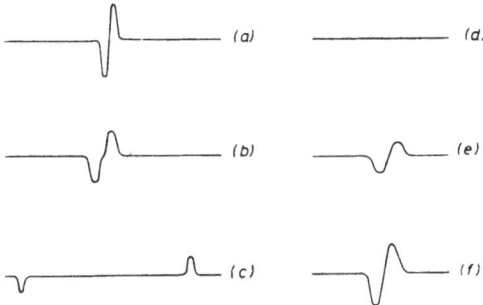

Fig. 3.30. Spectra observed from a single-line resonance using a double-modulation-recording spectrometer with the modulating frequency f_m much greater than the sample linewidth $(\Delta f)_E$. (a) to (c) Normal phasing of a phase-sensitive detector. (d) to (f) Reference phase shifted by 90°. (a) and (d) $f_m \ll \Delta f$. (b) and (e) $f_m \simeq (\Delta f)_E$. (c) and (f) $f_m \gg (\Delta f)_E$.

90° however, the central S-curve is observed and the sideband resonances are not. For normal operation, the single line splits into two when the modulation frequency is increased [see Fig. 3.30(c)]. The equivalent effect for the quadrature phase relationship is shown in Fig. 3.30(f). The general conclusion is that the modulating frequency should be kept low compared with the effective linewidth of the resonance line. According to equation (1.5), a value of 100 kc/s corresponds to a linewidth of 30 milligauss, which is adequate for most purposes.

Another factor determining the choice of modulation frequency f_m is that of enhanced a.m. klystron noise, which was introduced in §3.10 by reference to Fig. 3.19. In Fig. 3.19, the circle represents the variation of reflection coefficient with frequency for a matched reflection-cavity spectrometer. The expression from which this is derived is equation (3.8), with $\omega^2 M^2/R_0$ put equal to R. In this situation, and with the cavity reactance X expressed fully, the voltage V_L reaching the detector from the cavity is given by

$$V_L = \frac{-E_g}{2} \cdot \frac{j\left(\omega L - \frac{1}{\omega C}\right)}{2R + j\left(\omega L - \frac{1}{\omega C}\right)} \tag{3.40}$$

Thus, for resonators of high Q-factor, and using the definition of unloaded Q-factor Q_U developed in §3.4, V_L can be written for frequencies in the region of resonance as

$$V_L \simeq \frac{-E_g}{2} j Q_U \frac{\delta\omega}{\omega_0}\left(1 + j Q_U \frac{\delta\omega}{\omega_0}\right) \qquad (3.41)$$

Now from Fig. 2.24 the optimum drive power for a normal microwave detector is in the region of 0·5 mW. Thus, for a typical klystron power of, say, 50 mW, the overall attentuation of the bucking arm should be 20 dB, or a voltage loss of 1/10. Thus the ratio of the diameter of the circle to A − B, the drive vector in Fig. 3.19, is 10 : 1.

Now, for very low frequency noise modulation of the klystron, the noise sidebands are attenuated by the same factor, but as the modulating frequency increases, the attenuation is reduced as the sidebands begin to be reflected from the cavity. (The reflection of the sidebands is due to the cavity's finite reactance at the noise-sideband frequencies.) It is therefore reasonable to define a critical frequency ω_{mc} at which the reflection from the cavity is equal to the attenuation of the bucking arm. In general, ω_{mc} is obtained by putting the cavity-reflection coefficient, given by $2V_L/E_g$ from equation (3.41), equal to the bucking attenuation. For large values of bucking attenuation however, $Q_U \delta\omega/(\Delta\omega) \ll 1$, so that

$$V_L \simeq \frac{E_g}{2}\left(-j Q_U \frac{\delta\omega}{\omega_0}\right) \qquad (3.42)$$

Thus, for the conditions suggested, and for a typical unloaded Q-factor, Q_U, of 5000 and an operating frequency of 10 kMc/s, the critical frequency $f_{mc}(= \omega_c/2\pi)$ is equal to 200 kc/s. Thus 100 kc/s is again a good choice for the modulating frequency because it just avoids enhancing the klystron noise. In my experience, even klystrons that do not ordinarily contribute noticably to the overall noise level because of their a.m. noise, begin to do so if the modulating frequency exceeds this critical frequency. In particular, measurements at a frequency of 500 kc/s have confirmed this point.

§3.15

DRIFT AND $1/f$ NOISE

IN DOUBLE-MODULATION RECORDING

It was shown in §2.3 that, in the ordinary way, the signal-to noise ratio (S/N) for spectra traced at the output of a double-modulation-type spectrometer is proportional to the square root of the time T_r taken to record the spectra. If T_r is increased, the time constant $C_1 R_1$ (say T_f) of the low-pass filter following the phase-sensitive detector can be increased in proportion, and as the bandwidth of the system $(\Delta f)_f = (2\pi T_f)^{-1}$, the bandwidth of the spectrometer is accordingly reduced. Then, assuming the noise to be predominently white, the noise voltage is proportional to $(\Delta f)_f^{-\frac{1}{2}}$, so that

$$\frac{S}{N} \sim (\Delta f)_f^{-\frac{1}{2}} \sim T_r^{\frac{1}{2}} \tag{3.43}$$

It is not, however, always correct to assume that the noise is white, because, even when double modulation is used, some $1/f$ noise and drift are present at the output of the spectrometer. The sources of this drift and noise may be listed as:

(i) Drift and noise originating from the phase-sensitive detector and the d.c. amplifier that follows it.

(ii) Cavity background effects.

(iii) Modulation leakage.

Because the normal value for the bandwidth of the low-pass filter at the output of the spectrometer is in the region of 1 c/s, the noise originating from the phase-sensitive detector and d.c. amplifier is almost entirely $1/f$ noise over the frequency range which falls within the band-pass of the filter. Thus, although the gain of the a.c. amplifier preceding the phase-sensitive detector is generally high enough for the white noise originating from the amplifier and microwave detecting system to over-ride the noise from the phase-sensitive detector and d.c. amplifier, there is always some frequency below which the $1/f$ noise predominates. It was shown in §2.2 that, for a periodically recorded signal, the signal-to-noise ratio is not improved by increasing the time taken to make one recording. This is because an increase in the period of the signal results in a proportional reduction in

both the upper and lower cut-off frequencies of the system used
to amplify and display the signal. Then, because the noise is
$1/f$ noise, the increase in spectral intensity that the reduction in
mean frequency causes, exactly balances the improvement due
to the reduction in bandwidth. When spectra are recorded only
once, the same argument applies, because, although no network
introducing a lower cut-off frequency is present, one could be
without making any difference, provided that its cut-off frequency
is low compared with the inverse of the time taken to make the
recording. If the recording time is increased, the cut-off frequency
of the hypothetical network then needs to be lowered accordingly.
Thus, as the recording time T_r is increased, the $1/f$ noise from
the phase-sensitive detector and d.c. amplifier remains unaltered,
whilst the allowed reduction in bandwidth causes the white
noise from the earlier parts of the spectrometer to be reduced
according to equation (3.43). In addition, the effects of drift are
certainly not reduced by increasing the time of recording, so,
even if, for normal values of recording time, the white noise predominates as the time is increased, there eventually comes a
point where the white noise falls below the $1/f$ noise and drift,
and no further improvement is obtained.

In practice, the $1/f$ noise from the output of a normal double-
modulation spectrometer is usually a little below the legitimate
white noise for the commonly used values of T_f, which are in
the region of 1 sec. If, however, the recording time and filter
time constant are increased much beyond this range of values,
the $1/f$ noise soon begins to predominate, and little advantage is
gained. In fact, the phase-sensitive detector and d.c. amplifier
are seldom the major sources of drift and $1/f$ noise, and more
likely causes are the cavity background and modulation leakage
effects.

It is very common to find, when operating an E.S.R. spectrometer, that, when the field modulation is applied, some deflection of the pen recorder is obtained, even in the absence of a
sample. This deflection may simply be due to some of the output
from the modulator leaking into the a.c. amplifier, or even to the
reflector of the klystron. These effects are classed as modulation leakage and can usually be reduced to negligible proportions
by careful design. There often remains, however, when modulation

leakage has been cured, a spurious standing signal which appears to vary somewhat with the value of applied field from the magnet. This effect is termed '*cavity background*' and is still largely not understood. One point that has been investigated is the possibility that the applied field modulation is causing the cavity walls, or, where an internal loop is used, the modulating loop, to vibrate. There is little doubt that such vibration does occur, particularly at lower modulating frequencies and where a badly supported internal loop is used. However, the effect of the vibration is to modulate the resonant frequency of the cavity relative to the klystron. Hence, by accurately tuning the klystron to the cavity the effect should be avoided. This is partly true; the background signal is certainly minimized if the klystron is correctly tuned, but it is not eliminated. Moreover, normal precautions to stop vibration seldom eliminate the effect. As the chemical condition of the cavity wall influences the extent of the cavity background, it seems probable that the effect is partly due to vibration and partly to some kind of magnetic resonance in the surface of the wall. Whatever the cause the effect is seldom absent and unfortunately rarely constant with time, even at a constant field value. In fact the fluctuations in the cavity background are probably $1/f$ dependent, as are usually the fluctuations in modulation leakage if present.

§3. 16

SECOND-DERIVATIVE RECORDING

One way of avoiding the effects of $1/f$ noise and drift at the output of a double-modulation spectrometer is simply to apply the technique of double modulation once again. Starting with a conventional double-modulation system, a further sinusoidal modulation is applied to the main magnetic field (see Fig. 3. 31). If the amplitude of the second modulating signal is small compared with the width of the magnetic resonance line, the effect of the modulation is to superimpose on the spectra traced on the recorder an a.c. signal of an amplitude proportional to the derivative of the spectra. Then, instead of feeding the signal to the recorder, the signal is fed to an amplifier which is tuned to the frequency of the field modulation and so only accepts the

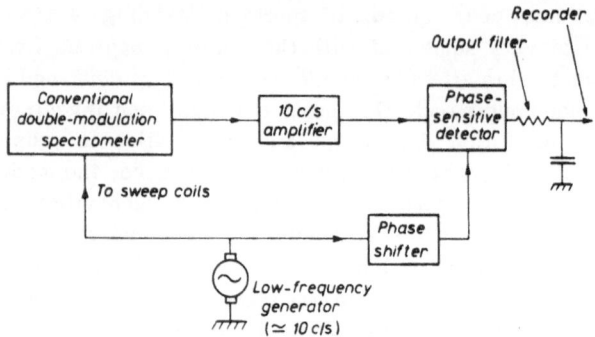

Fig. 3.31. Second-derivative recording system.

derivative signal. The output from the amplifier is then fed to a phase-sensitive detector and a low-pass filter before being recorded.

With this arrangement, the a.c. amplifier only accepts noise from the output of the original spectrometer which is centred around the second modulating frequency. Thus, provided the second modulating frequency is above the value at which $1/f$ noise becomes significant at the output of the original system, $1/f$ noise and drift are avoided. Further reduction of the bandwidth of the system is obtained by reducing the time constant of the low-pass filter following the second phase-sensitive detector, and, according to §2.3, this merely has the effect of narrowing the range of noise components accepted from the original system. The central frequency remains at the value of the second modulating frequency. Since the frequency at which $1/f$ noise becomes significant is usually about $1\,c/s$, a value of $10\,c/s$ is suitable for the second modulating frequency. This frequency is fortunately low enough to present no problems of penetration into the cavity. The modulation can therefore always be applied by coils placed outside the cavity.

A possible disadvantage of the above system is that the final display is proportional to the second derivative of the magnetic resonance spectrum. This is not always an objection however, because the position of the line is defined, as shown in Fig. 3.32, by a peak rather than a zero. The problem of resolving complex spectra is then sometimes made easier.

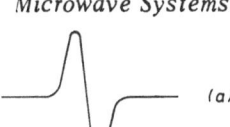

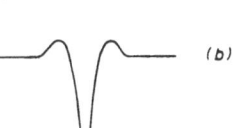

Fig. 3.32. The first (*a*) and second (*b*) derivatives of an E.S.R.
absorption line.

§3.17

COMPUTER OF AVERAGE TRANSIENTS

A more elegant method of overcoming the $1/f$ noise at the output of a spectrometer is to use a computer of average transients (C.A.T.).[8] This device records a trace fairly rapidly which it stores in its memory. A second trace is then recorded, and the computer adds it to the first. This process is then repeated for many subsequent traces and results in an improvement in the signal-to-noise ratio of the stored 'sum' trace, which can be monitored continuously. An improvement in the signal-to-noise ratio results because the spectrum in each trace is exactly the same, so the result of n measurements is to produce a spectrum n times as high as the original. The noise components on the various traces are, however, completely uncorrelated, so that the final noise amplitude is multiplied by $\sqrt{n}$. Thus an improvement in the signal-to-noise ratio of $\sqrt{n}$ is obtained. Now equation (3.43) indicated that if, using a conventional spectrometer, the recording time is multiplied by n, the gain in signal-to-noise ratio is again equal to $\sqrt{n}$. Thus the particular mode of operation makes no fundamental difference to the signal-to-noise ratio obtained, provide that the $1/f$ noise is avoided. Hence, because second-derivative recording and the C.A.T. are both successful means of avoiding the $1/f$ noise, they can be used with equal advantage to allow the recording time to be usefully extended.

The C.A.T. does however have several important subsidiary advantages. First, in normal and second-derivative modes the length of recording must be prejudged, whilst with the C.A.T. the experiment can be stopped as soon as an adequate signal-to-noise ratio is obtained. Second, the effect of drift in the magnetic field can be reduced. If a proton resonance field marker (see §8.6) is provided, this can be used to start the sweep of the C.A.T., and so the C.A.T. always begins to record at the same value of field, irrespective of long-term drift. The final advantage is simply that the first, rather than the second, derivative is displayed. Against these advantages the cheapness of the second-derivative technique must be considered.

The method of operation of the C.A.T. is to sample the spectra at successive intervals after the beginning of the linear sweep, and to convert each reading to a digital number which is stored in successive sections of a digital memory store. Once the first sweep has been completed and the second begun, the samples from the first sweep are successively extracted from the memory, added to the second and replaced. In fact, the full process of extraction, addition and replacement takes a minimum of about $25\,\mu s$, so for 1024 ($= 2^{10}$) channels a full sweep takes about 25 ms. Thus a sweep frequency as high as 40 c/s can be used, which is just enough to avoid the $1/f$ noise, bearing in mind that the system is equivalent to having a low-frequency cut-off of about 0·4 c/s. A C.A.T. is invaluable for improving the signal-to-noise ratio for short-lived paramagnetic materials generated by reactions that can be repeated periodically.

§3.18

BACKWARD DIODE DETECTORS

Recently a new type of diode has been developed where $1/f$ noise has been greatly reduced. The diode is known as a back-diode. It gives, as can be seen from Fig. 3.33(a), a typical variation in noise figure with operating frequency. The corresponding curve for a normal diode is shown in Fig. 3.33(b). Thus, using back-diodes, it might appear that there need be no compromise at all and that the modulating frequency could be lowered to 10 c/s. This would give a resolution of 3 milligauss, which is nearly always adequate. However, the reduction in noise is partly due to the discontinuity in the static characteristic of the diode (§2.7)

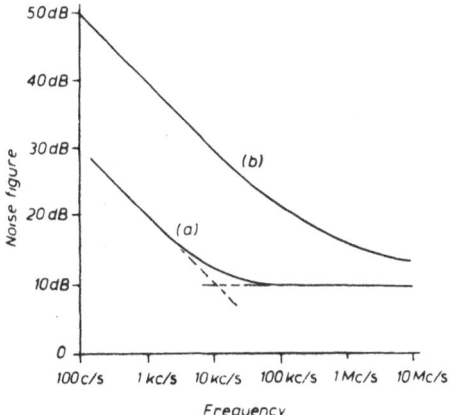

Fig.3.33. Variation of noise figure with frequency of a typical backward diode microwave mixer (a) compared with that for a conventional diode (b).

being much sharper, and so less local oscillator drive is required to obtain the maximum conversion gain and thus the optimum noise figure. The bucking arm attenuation is therefore greater and this reduces the frequency [defined by equation (3.43)] at which the klystron a.m. noise begins to be exalted. The sensitivity to f.m. noise for a given degree of off-tuning is also greater. These are factors that have yet to be investigated and may not be very troublesome. If not, there is little doubt that the back-diode will tend to be used more frequently in E.S.R. spectrometers and may render the superheterodyne of Chapter 5 unnecessary for all but specialized applications.

References

1. Wilmshurst, T.H., Gambling, W.A., and Ingram, D.J.E., *J. Electron. Con.*, 1962, **13**, 339.
2. Feher, G., *Bell Syst. tech. J.*, 1957, **36**, 449.
3. Frazer, W., *Telecommunications* (MacDonald, 1957).
4. Faulkner, E.A., *J. sci. Instrum.*, 1962, **39**, 135.
5. Henning, J.C.M., *Rev. sci. Instrum.*, 1961, **32**, 35.
6. Mehlkopf, A.F., and Smidt, J., *Proc. XI Colloque Ampère*, 1962.
7. Teaney, D.T., Klein, M.P., and Portis, A.M., *Rev. sci. Instrum.*, 1961, **32**, 721.
8. Klein, M.P., and Barton G.W., *Rev. sci. Instrum.*, 1963, **34**, 754.

Chapter 4

Spectrometer Cavities

In Chapter 2 it is shown that the effect of placing a sample exhibiting .E.S.R. in the cavity of an E.S.R. spectrometer is to couple an effective spin-generator in series with the equivalent circuit of the cavity. For an unsaturable sample, the criterion of sensitivity for the cavity is that the available power from the spin-generator should be as high as possible for a given power dissipated in the cavity by the klystron. In this chapter the corresponding criterion for saturable samples is deduced. Both criteria are then expressed mathematically and considered in some detail for various types of sample. In particular, the effect of the frequency of operation and the limitations to sample size imposed by the inhomogeneity of the magnet are taken into consideration. The chapter includes a short account of the practical considerations of spectrometer cavity design, and concludes with a somewhat speculative consideration of the possible replacement of the cavity by a slow-wave structure.

§4.1

SENSITIVITY EQUATIONS

Using the equivalent circuit for the E.S.R. cavity discussed in §2.5 and shown in Fig. 4.1, we shall now relate the available power P_a from the sample spin-generator to the power P_i dissipated in the cavity by the klystron. L and C of Fig. 4.1 are assumed to resonate at the klystron frequency ω and their combined impedance is therefore zero. R represents the losses within the cavity. E_R and E_X represent the two components of the sample spin-generator voltage, E_R being the resistive or absorbing component and E_X the reactive or dispersive component. Then, since the available power P_a from the two components is

108

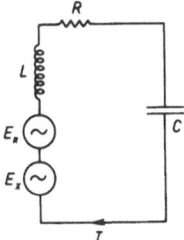

Fig. 4.1. Equivalent circuit of an E.S.R. spectrometer cavity containing a sample exhibiting rotating magnetization.

equal to $(E_R^2/8R)$ and $(E_X^2/8R)$, it becomes necessary to determine E_R and E_X in terms of P_i. In the equivalent circuit, however, P_i is directly related to the current I due to the klystron generator, because $P_i = I^2R/2$. Thus the problem resolves to that of relating E_R and E_X to I. Now $E_R I/2$ represents the power absorption of the sample, which can also be determined in terms of the interaction of the rotating magnetization of the sample with the field of the cavity. Thus it becomes directly possible to obtain E_R, and E_X can then be obtained from the relationship between absorption and dispersion given by equations (2.13) and (2.14).

Consider, therefore, the small contribution δE_R to E_R from an element of sample of volume δv. The power δP absorbed by the element of sample is then equal to $I\,\delta E_R/2$. In §2.5 it is shown that, at the centre of the magnetic resonance line, the magnetic moment $\delta \mu_+$ from the element rotates at an angle of $90°$ behind the applied rotating field H_+. The torque between a magnetic moment μ and a magnetic field H is given by $\mu \wedge H$. Thus, at the centre of the line, the power fed from the field H_+ to the magnetic moment $\delta \mu_+$ is equal to $\omega \delta \mu_+ H_+$. This power is equal to δP, so that

$$\delta E_R = \frac{2\omega \delta \mu_+ H_+}{I}$$

Now $\delta \mu_+ = M_+ \delta v$, where M_+ is the intensity of magnetization of the sample at the point of the volume element. Thus, integrating over the entire sample to find E_R, one has

$$E_R = \frac{2\omega}{I} \int_s M_+ H_+ dv$$

As the available power P_a from the spin-generator is equal to $E_R^2/8R$,

$$P_a = \frac{\left(\omega \int_s M_+ H_+ dv\right)^2}{2I^2R}$$

But $I^2R/2$ is the power P_i dissipated in the cavity by the generator, so

$$P_a = \frac{\omega^2}{4}\left(\frac{\int_s M_+ H_+ dv}{P_i}\right)^2 P_i \qquad (4.1)$$

Equation (4.1) is expressed in terms of rotating components of the field and magnetization, which are not appropriate to the practical case. If, however, H is the linearly polarized field, of which H_+ is a component, then $H_+ = H/2$. Now the rotating magnetization M_+ gives a linearly polarized component M in the direction of the applied field so that $M = H_+$. Inserting these relations into equation (4.1) gives

$$P_a = \frac{\omega^2}{16}\left(\frac{\int_s MH dv}{P_i}\right)^2 P_i \qquad (4.2)$$

At the centre of the sample resonance, $M = \chi''H$, so that

$$P_a^{\frac{1}{2}} = \frac{\omega\chi''}{4}\frac{\int_s H^2 dv}{P_i} P_i^{\frac{1}{2}} \qquad (4.3)$$

Consequently, for unsaturable samples of a given susceptibility and of a given operating frequency ω, the cavity is simply required to concentrate the greatest strength of field over the largest possible volume of sample for a given power dissipated in the cavity by the klystron. The reason for expressing equation (4.3) in terms of $P_a^{\frac{1}{2}}$ is that the voltage developed by the sample at the detector is proportional to $P_a^{\frac{1}{2}}$, and so is the amplitude of the signal traced on the recorder at the output of the spectrometer.

To conform with most existing treatments it should be pointed out here that the term $\omega \int_s H^2 dv/P_i$ is proportional to the product of the Q-factor of the cavity and of the cavity filling factor η. The basic definition of the Q-factor of a resonator is, from §3.4,

$$Q = \frac{\omega \times \text{Energy stored}}{\text{Power dissipated}} \tag{4.4}$$

When the magnetic field is at its peak, all the energy stored in the cavity is in the magnetic field. The energy stored in the magnetic field[1] of the cavity at this instant is then given by $(\mu_0/2)\int_c H^2 dv$, where $\int_c dv$ represents the integral over the volume of the cavity. Thus from (4.4)

$$Q = \omega\mu_0 \frac{\int_c H^2 dv}{2P_i} \tag{4.5}$$

and combining (4.3) and (4.5) gives

$$P_a^{\frac{1}{2}} = \frac{Q\chi''}{2\mu_0} \cdot \frac{\int_s H^2 dv}{\int_c H^2 dv} P_i^{\frac{1}{2}} \tag{4.6}$$

Here $\int_s H^2 dv / \int_c H^2 dv$ is the filling factor η, which indicates how efficiently the sample fills the magnetic field available in the cavity. When the sample fills the entire cavity, the value of η is unity. Thus,

$$P_a^{\frac{1}{2}} = \frac{\chi'' Q \eta}{2\mu_0} P_i^{\frac{1}{2}} \tag{4.7}$$

Equation (4.7) is a commonly used form for expressing the sensitivity of the cavity, because η, the filling factor, is quite easy to visualize, whilst the Q-factor is a well-established parameter which can be determined from the cavity bandwidth (see §3.4). However, as a fundamental criterion for determining the cavity sensitivity, equation (4.7) is less direct than equation (4.3), and so in the following discussion the latter will be used in preference to the former.

When it is possible to saturate the sample, equation (4.3) becomes less useful. Then P_i no longer remains constant, but is adjusted, as the cavity system is altered, so that saturation of the sample is just avoided. In fact, this demands that the value of magnetization M_{max} at the point of most intense magnetization within the sample is limited to the saturation microwave magnetization M_{sat} of the sample. M_{sat} here is defined as the maximum microwave magnetization obtainable without saturation. Then, because the ratio of magnetization M to field H is constant for varying positions over the sample, equation (4.2) may be written as

$$P_a^{1/2} = \frac{\omega}{4}\left(\int_s M^2 dv \frac{\int_s H^2 dv}{P_i} \right)^{1/2} \tag{4.8}$$

Then, $\int_s M^2 dv$ can be expressed as $M_{max} \int_s \hbar^2 dv$, where $\hbar$ is a dimensionless function which has the same spatial variation as M, and therefore also of H, and a value of unity at the point of M_{max}. Equation (4.8) then becomes

$$P_a^{1/2} = \frac{\omega}{4}\left(\int_s H^2 dv \frac{\int_s \hbar^2 dv}{P_i} \right)^{1/2} M_{sat} \tag{4.9}$$

where M_{max} is put equal to M_{sat}.

The term $\int_s H^2 dv/P_i$ is thus exactly as before, but the additional factor of $\int_s \hbar^2 dv$ indicates that whilst, for a given dissipated power, it is still necessary to concentrate the largest possible field over the greatest volume of sample, greater importance must be given to the fact that the field should extend over a large volume of sample.

As the power dissipated in the cavity is of importance, it is necessary to consider the factors contributing to the cavity losses. These will consist of ohmic losses in the cavity wall and dielectric losses in the sample. The power loss in the cavity walls is given by $(R_s/2)\int_w H^2 da$, where R_s is the surface resistivity of the wall and $\int_w H^2 da$ the surface integral of the magnetic field squared over the entire area of the inside cavity walls. The power loss in the sample is given by $1/2.\omega\epsilon_0\epsilon''\int_s E^2 dv$, where ϵ'' is the imaginary part of the complex permittivity and $\int_s E^2 dv$ the volume integral of the square of the electric field strength over the sample. Expressed thus, equations (4.3) and (4.9) for the sensitivity of the cavity for unsaturable and saturable samples become respectively

$$P_a^{1/2} = \frac{\omega \chi''}{2} \cdot \frac{\int_s H^2 dv}{R_s \int_w H^2 da + \omega \epsilon_0 \epsilon'' \int_s E^2 dv} \cdot P_i^{1/2} \qquad (4.10)$$

and

$$P_a^{1/2} = \frac{\omega}{4} \cdot \left(\frac{2\int_s H^2 dv \cdot \int_s h^2 dv}{R_s \int_w H^2 da + \omega \epsilon_0 \epsilon'' \int_s E^2 dv} \right)^{1/2} \cdot M_{sat} \qquad (4.11)$$

§4.2

FREQUENCY DEPENDENCE OF CAVITY SENSITIVITY

From equations (4.10) and (4.11) it is clear that the operating frequency ω is one factor influencing the sensitivity of an E.S.R. spectrometer cavity. However, apart from entering the expressions explicitly, the parameters χ'', M_{sat}, R_s and ϵ'' are also all frequency dependent.

By putting $\delta\omega = 0$ in equation (2.14), we obtain the value of χ'' at resonance

$$\chi'' = \frac{-\pi T_2 g \beta M_z}{h} \qquad (4.12)$$

Here M_z is the component of magnetization in the direction of the magnetic field and is equal to the unperturbed value M_0 for microwave field values well below saturation. M_0 can be determined from the Boltzmann distribution of equation (1.6). If, for a two-level system, N is the total number of spins per unit volume and ΔN is the excess population of the lower level $N_1 - N_2$, then ΔN, the net number of spins per unit volume aligned with the applied steady magnetic field, is given by

$$\Delta N = N \tanh \frac{hf}{2kT} \qquad (4.13)$$

As $hf \ll kT$, equation (4.13) gives for microwave frequencies

$$\Delta N \simeq \frac{Nhf}{2kT} \qquad (4.14)$$

Now, for a free electron the effective magnetic moment is β and for a bound electron $g\beta/2$. Thus the intensity of magnetization, i.e. the magnetic moment per unit volume, M_0, in the direction of the applied field is given by

$$M_0 = \frac{g\beta Nhf}{4kT} \tag{4.15}$$

Inserting equation (4.15) into equation (4.12) and putting $M_z = M_0$ we obtain for negligible saturation

$$\chi'' = \frac{-\pi T_2 g^2 \beta^2 Nf}{4kT} \tag{4.16}$$

Therefore χ'' is proportional to frequency.

For saturable samples, equation (4.11) gives the criterion of sensitivity and here the term M_{sat} broadly replaces χ'' in the previous equation. The microwave susceptibility χ'' links M_{sat} to the limiting value of applied microwave magnetic field H_{sat}, because

$$M_{sat} = \chi'' H_{sat} \tag{4.17}$$

Now the low-field value of χ'', given by equation (4.16), is obtained by putting in equation (4.12) M_z equal to M_0. More generally, M_z is given by

$$M_z = M_0 \left[1 + T_1 T_2 \left(\frac{\pi g\beta H}{h} \right)^2 \right]^{-1} \tag{4.18}$$

Thus, to avoid saturation H must be small compared with $h/\pi g\beta(T_1 T_2)^{1/2}$. This limitation to H is independent of frequency, so, from equation (4.17), M_{sat} is dependent only on the frequency dependence of χ'' and is therefore again directly proportional to frequency. Combining equations (4.16) and (4.17) one has

$$M_{sat} = \frac{\pi T_2 g^2 \beta^2 NfH_{sat}}{4kT} \tag{4.19}$$

From Ramo and Whinnery[1] the surface resistivity R_s of the cavity wall is given by

$$R_s = \left(\frac{\pi f\mu}{\sigma} \right)^{1/2} \tag{4.20}$$

where μ is the permeability of the material and σ its conductivity. Consequently R_s is inversely proportional to the square root of the frequency. To express fully the frequency sensitivity, equations (4.10) and (4.11) can, for unsaturable samples, now be written in the form

$$P_a^{\frac{1}{2}} = \frac{K_u f^2 \int_s H^2 dv . P_i^{\frac{1}{2}}}{\left(\frac{\pi\mu}{\sigma}\right)^{\frac{1}{2}} f^{\frac{1}{2}} \int_w H^2 da + 2\pi f \epsilon_0 \epsilon'' \int_s E^2 dv} \qquad (4.21)$$

and for saturable samples

$$P_a^{\frac{1}{2}} = K_s f^2 \left[\frac{\int_s H^2 dv . \int_s R^2 dv}{\left(\frac{\pi\mu}{\sigma}\right)^{\frac{1}{2}} f^{\frac{1}{2}} \int_w H^2 da + 2\pi f \epsilon_0 \epsilon'' \int_s E^2 dv}\right]^{\frac{1}{2}} H_{sat} \qquad (4.22)$$

where $K_u = \pi^2 T_2 g^2 \beta^2 N/4kT$ and $K_s = K_u/\sqrt{2}$.

The next step is to find the sample and cavity geometry which gives the optimum sensitivity as expressed by equations (4.21) and (4.22). In general, this is a somewhat complicated problem and it proves convenient to consider the extreme cases of samples with negligible dielectric loss, where $\epsilon'' = 0$, and samples with severe dielectric loss, such as water, where ϵ'' is high. Intermediate cases can then be treated easily by interpolating between the two results. It is also necessary to distinguish between samples of unlimited availability and very small samples of fixed size. Furthermore, the small fixed samples are usually single crystals which cannot be grown above a certain size and which are seldom very lossy. These samples, on account of their small size, can be confined fairly well to the region of zero electric field within the cavity. Thus the small sample need only be considered for the case of zero dielectric loss. We shall therefore consider the following classes of sample: (*a*) samples of unlimited availability and zero dielectric loss, (*b*) samples of unlimited availability and severe dielectric loss, and (*c*) small samples of fixed size whose dielectric loss can be neglected.

§4.3

SAMPLES OF UNLIMITED AVAILABILITY AND ZERO DIELECTRIC LOSS

For samples of no dielectric loss, the equations giving the sensitivity of the cavity are for unsaturable and saturable samples respectively

$$P_a^{1/2} = K_u f^{3/2} \frac{\int_s \hbar^2 dv}{\left(\frac{\pi\mu}{\sigma}\right)^{1/2} \int_w \hbar^2 da} P_i^{1/2} \qquad (4.23)$$

$$P_a^{1/2} = K_s f^{7/4} \frac{\int_s \hbar^2 dv}{\left(\frac{\pi\mu}{\sigma}\right)^{1/4} (\int_w \hbar^2 da)^{1/2}} H_{sat} \qquad (4.24)$$

Thus, apart from the obvious requirement of a high conductivity for the cavity wall, the problem resolves in each case to obtaining a high ratio of $\int_s \hbar^2 dv$ to $\int_w \hbar^2 da$ and a high value of f. For a given sample and a given operating frequency, therefore, the requirement of the cavity is that it should concentrate the field over a large volume of sample, whilst allowing the minimum area of cavity wall to come into contact with the field. In other words, the cavity should 'fit' the sample as closely as possible. It is also possible to give a very crude interpretation to the ratio $\int_s \hbar^2 dv / \int_w \hbar^2 da$ by assuming that $\hbar$ either is constant or oscillates in some manner over the walls of the cavity and the sample. Therefore, $\int_s \hbar^2 dv$ and $\int_w \hbar^2 da$ are, respectively, equal to about half the volume of the sample and to about half the area of the cavity wall. Thus the ratio $\int_s \hbar^2 dv / \int_w \hbar^2 da$ is roughly equal to the ratio between the volume of the sample and the area of the cavity wall. Besides the sample having to fill as much of the cavity as possible, the cavity and sample should be large and of uniform proportions. A long and thin cavity, for example, would give an unnecessarily high ratio of area to volume. If we make the above approximations and call the linear dimension of the cavity and sample $2x_0$, from equations (4.23) and (4.24) we obtain for the unsaturable sample

$$P_a^{1/2} \simeq 2K_u \left(\frac{\sigma}{\pi\mu}\right)^{1/2} f^{3/2} x_0 P_i^{1/2} \qquad (4.25)$$

and for the saturable sample

$$P_a^{1/2} \simeq 4K_s \left(\frac{\sigma}{\pi\mu}\right)^{1/4} f^{7/4} x_0^2 H_{sat} \qquad (4.26)$$

Thus it is clear that x_0 should in both cases be as large as possible. However, the sample size is restricted by the proportions of the magnet and this restriction depends on the value of the magnetic field, which is dependent on the operating frequency [see equation (1.5)]. Hence, in order to determine the optimum operating frequency, it is necessary to find the frequency dependence of the maximum allowable value of x_0.

The ratio of the size of the sample to the size of the magnet is determined primarily by the field homogeneity required over the sample, whilst the size of magnet is limited by factors such as weight, cost, electrical driving power and saturation magnetization of the material from which the magnet is constructed. The magnetic field needs to be homogenious over the sample, because, if the field varies from one part of the sample to another, different regions of the sample resonate at different values of current through the coils of the magnet. This effect tends to broaden the resonance line and it is therefore clear that the inhomogeneity over the sample must be small compared with the natural linewidth of the magnetic resonance. It is usual to express the inhomogeneity of the field within the pole gap of a magnet by writing the field as a series of spherical harmonics.[2] When the ratio of the pole gap S_P to the pole diameter D_P is small, this approach resolves into expressing the factional variation $\delta H/H$ in field from its maximum value as a polynomial in r, the distance from the centre of the pole gap. Thus,

$$\frac{\delta H}{H} = a_2 \left(\frac{S_P}{D_P}\right)^3 \left(\frac{r}{S_P}\right)^2 + a_4 \left(\frac{S_P}{D_P}\right)^5 \left(\frac{r}{S_P}\right)^4 + a_6 \left(\frac{S_P}{D_P}\right)^7 \left(\frac{r}{S_P}\right)^6 + \cdots$$

$$(4.27)$$

where a_2, a_4, ..., are constants of the order of unity but which are somewhat greater for variations of r along the axis of the magnet than they are for variations of r in the plane of the pole face. Invariably therefore, the larger the pole diameter D_P, the larger the sample size $2x_0$, which corresponds to the maximum value of $2r$. However, the largest practicable size of pole diameter is about 12 inches and even this is not possible at all values of operating frequency. For field values of up to 3 kilogauss, corresponding to operating frequencies of up to 10 kMc/s [see equation (1.5)], this size is quite possible, but for higher frequencies requiring higher field values, the power required to drive the magnet becomes somewhat excessive. Beyond about 10 kilogauss the material of which the magnet is made tends to saturate. Thus, to obtain fields of 10 kilogauss and above, it is necessary to operate the yoke of the magnet near to saturation and to concentrate the available flux by using high coercivity, coned pole pieces. This technique is feasible for fields up to about 18 kilogauss, that is to frequencies of 60 kMc/s, but beyond this point normal magnets cannot be easily used.

It is difficult to give, for the range 3 to 18 kilogauss (10 to 60 kMc/s), any completely satisfactory expression relating the pole diameter D_P to the frequency f, but a purely empirical law of $D_P \sim f^{-1}$ is not too erroneous. Then, assuming $x_0 \sim D_P$, equations (4.25) and (4.26) give for unsaturable samples

$$P_a^{\frac{1}{2}} \sim f^{\frac{1}{2}} \tag{4.28}$$

and for saturable samples

$$P_a^{\frac{1}{2}} \sim f^{-\frac{1}{4}} \tag{4.29}$$

Thus, for saturable samples, it seems advisable to use the lower frequency, whilst for unsaturable samples the sensitivity increases with frequency. It is not, however, always correct to assume that the sample size is proportional to the size of the magnet. Generally speaking, the thickness of the cavity wall and of the sample container is more or less fixed, so that the sample size x_0 tends to decrease more rapidly than the magnet size, and the advantages of using the higher frequency are very much in question even with unsaturable samples. Thus, for unlimited availability of sample, it is usually best to operate at the lower end of the frequency range of 10 to 60 kMc/s.

If the frequency is reduced further, the sample size becomes largely fixed for a given homogeneity, because it is not practical to increase the size of magnet. Thus x_0 is fixed in equations (4.25) and (4.26), and the frequency dependence is such that the highest possible frequency should be used for both saturable and unsaturable samples. Thus the most common operating frequency for an E.S.R. spectrometer is, in fact, 10 kMc/s, and other frequencies are only for more specialized types of sample.

§4.4

PRACTICAL CAVITY DESIGN

In §4.3 we found that the best operating frequency for lossless samples of unlimited availability is 10 kMc/s. Therefore, all we have to do is to design a cavity for use at this frequency, bearing in mind that the cavity must fit the sample as closely as possible and also allow the largest possible amount of sample to be used. Now the sample dimensions are fundamentally restricted by the homogeneity required of the magnet, and so samples with narrow linewidths need to be smaller than those with wide lines. However, it is clear from equation (4.27) that, for a given degree $\delta H/H$ of inhomogeneity, the sample size, $2x_0 = 2r$, increases with decreasing pole gap S_P. Therefore the sample should fill the pole gap as far as possible and, if allowance is made for the thickness of the cavity wall, r/S_P in (4.27) may be taken as equal to 1/4. Then approximate values of $\delta H/H$ for a field of 3 kilogauss are 25 gauss for $S_P/D_P = 1/2$, 0·75 gauss for $S_P/D_P = 1/4$ and 25 milligauss for $S_P/D_P = 1/6$. Thus, for a 12-inch pole face, an average figure for the pole gap is 2 inches $(S_P/D_P = 1/6)$. The size of cavity is then of the same order as the wavelength corresponding to the frequency of operation, and hence the cavity systems of Figs. 4.2(a) and (b) are reasonably appropriate. These are the H_{012} rectangular and H_{011} cylindrical modes, which have resonant frequencies given[1] by

$$f = \left[\left(\frac{c}{2b}\right)^2 + \left(\frac{c}{d}\right)^2\right]^{1/2} \tag{4.30}$$

and

$$f = \left[\left(\frac{c}{1 \cdot 64a}\right)^2 + \left(\frac{c}{2d}\right)^2\right]^{1/2} \tag{4.31}$$

respectively, where c is the velocity of light. Both cavities are chosen because the maximum magnetic field runs down their centres and so the sample can be placed in the centre of the cavity, even when a sample tube is necessary. For a given sample size, the value of $\int_s h^2\,dv/\int_w h^2\,da$ turns out to be approximately the same for both, and, in fact, there is little to choose between the two. On the whole, the rectangular cavity is pre-

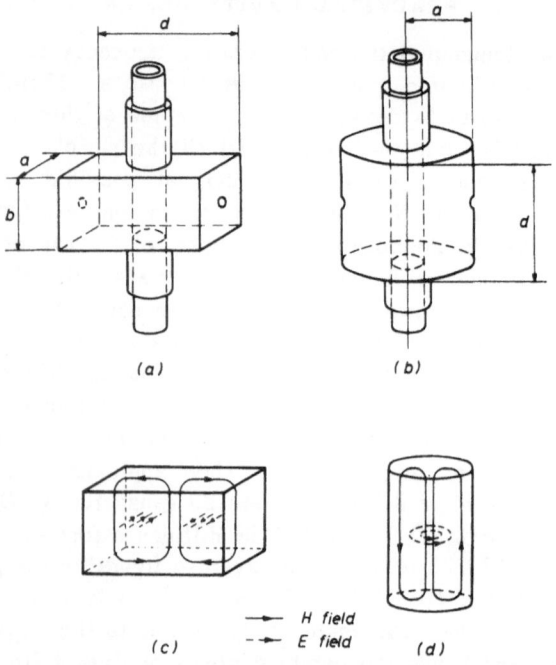

Fig. 4.2. Typical E.S.R. cavity configurations. (a) H_{012} rectangular cavity showing sample tube and coupling holes. (b) H_{011} cylindrical cavity showing sample tube and coupling holes. (c) Mode patterns in an H_{012} rectangular cavity. (d) Mode patterns in an H_{011} cylindrical cavity.

ferred for general applications, because it allows a closer pole spacing and therefore a better magnet homogeneity. The circular cavity is more often used for single-crystal studies, when it is desirable to be able to rotate the crystal relative to the magnetic field.

It is sometimes suggested that the cavity wall area could be reduced by using the H_{011} rectangular cavity configuration rather than the H_{012} configuration. This proposal, however, raises problems with regard to the sample tube, which, since the regions of maximum magnetic field in the H_{011} cavity are always in contact with a cavity wall, tends to hold the sample away from the region of maximum magnetic field.

The coupling to either cavity is generally via a coupling hole and the positions shown are typical but not invariable. For the transmission-cavity spectrometer both coupling holes are used, whilst for the reflection-cavity system one hole is blanked off. The metal pipes or 'stacks' connected to the holes through which samples are inserted into the cavity, are used to prevent power from being radiated out of the holes. The pipe diameter is carefully chosen so that the cut-off frequency of the circular waveguide is well above the frequency of the cavity. Only evanescent modes are then propogated up the stack, so, provided the length of the stack is greater than one or two wavelengths, no appreciable energy is lost. In calculating the cut-off frequency of the stack allowance must be made for a reduction in cut-off frequency caused by the dielectric constant of the sample tube and other glassware. If a conductor, such as the field modulating loop mentioned in §2.3, is placed down the stack, it is possible for T E M waves to propagate however small the diameter of the stack. Where such a loop is used, it is better to insert the wires through separate well-choked holes.

Sometimes a cavity has to be dismantled for cleaning, and when making provision for this requirement, it is preferable that joins in the cavity occur along the current lines (see Fig. 4.3).

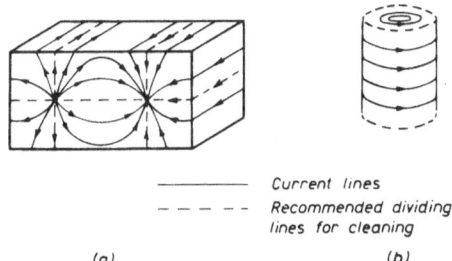

——————— *Current lines*
— — — — *Recommended dividing lines for cleaning*

(a)   (b)

Fig. 4.3. Surface current distributions in (a) an H_{012} rectangular E.S.R. cavity and (b) an H_{011} circular E.S.R. cavity.

For the rectangular cavity, the common practice of making the ends removable is therefore undesirable and it is better to make the breaks along the lines shown. For the circular cavity, the ends can be removed without cutting any current lines.

Often it is required to irradiate the sample when located in the cavity. This can be arranged by cutting slots along the lines shown for the rectangular and circular cavities. Alternatively, for the rectangular cavity, it is possible to cut a large window in the end and to fasten to it a section of waveguide reduced in

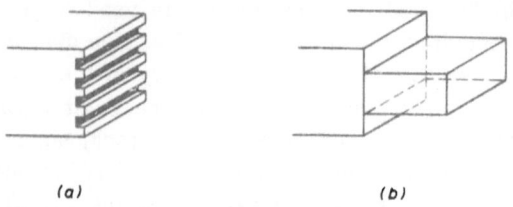

(a)   (b)

Fig. 4.4. Methods of allowing *in situ* irradiation of sample. (a) Slotted end wall. (b) Reduced height waveguide radiation port.

height, as in Fig. 4.4(b). This waveguide must be below the cut-off value for the frequency in use otherwise the waves propagate beyond the waveguide and energy is lost.

The problem of arranging for the modulating field to penetrate into the cavity has been approached in a variety of ways. One method is to make the cavity out of an insulating material and to plate chemically the inside with silver or some other suitable conductor. Pyrex glass[3] and epoxy resin[4,5] have been used successfully; hydrous aluminium silicate[6], which is initially soft and can be machined, becomes very hard after firing at 1000°C. Cook *et al.*[7] have obtained an improvement in the penetration of the modulating field by using an H_{011} circular cavity and machining a spiral groove along the whole length of the cylindrical wall. It is clear from Fig. 4.3(b) that this groove does not intercept the wall current and so the cavity Q-factor is not affected. A further advantage of this method is that it suppresses the E_{111} mode, which is normally degenerate with the H_{011} mode. The layer of conducting material in all but the last method must be thin compared with the skin depth of the modulating frequency and thick compared with the skin depth at the microwave frequency. It should perhaps be pointed out that the thickness *must* be several times the skin depth at the microwave frequency. The decay of current I from the value I_0 at the surface is exponential,[1] so that

$$I = I_0 \exp\left(-\frac{x}{\delta}\right) \tag{4.32}$$

where x is the thickness of the conducting material and δ is the skin depth. Now, if one considers the wave to be formed in the cavity by the standing wave energy bouncing between the walls once every cycle, then the ratio of energy radiated per cycle to energy stored is $\exp(-2x/\delta)$, because the energy $\sim H^2$ and $H \sim I$. The effective Q-factor due to the wall loss is then Q_{eff}, where

$$Q_{\text{eff}} = \frac{\omega \times \text{Energy stored}}{\text{Power radiated}}$$

which is equal to the above ratio $\times 2\pi$. Thus, to avoid degradation of the cavity Q-factor $\exp(2x/\delta) \gg Q/2\pi$. Actually, a value of $x = 5\delta$ gives a Q-factor of about 120,000, which represents a negligible loss compared with the normal losses of a cavity. Since, for most normal conductors at 10 Gc/s, $\delta \sim f^{-\frac{1}{2}}$, is equal to 10^{-3} mm and hence is 0·3 mm at 100 kc/s, a depth of 10^{-2} mm would seem a good compromise, but we must bear in mind that the above argument is based on only one wall being thin.

Unfortunately epoxy resins tend to crack and the chemical plating processes are not all that easy to apply. Thus manufacturers of commercial spectrometers tend to make the bulk of the cavity out of copper or copper-plated non-magnetic brass and add windows of quartz, which can, by evaporation, be conveniently coated with copper or silver. The best position to cut the windows is along lines of zero current, as shown in Fig. 4.5(a).

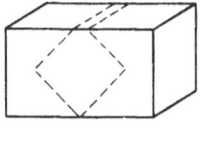 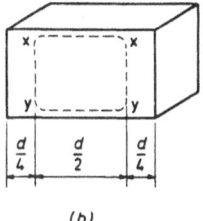

(a)   (b)

Fig. 4.5. Methods of inserting thin-wall windows into E.S.R. cavities. (a) The ideal method, and (b) a compromise method.

It is then important to place a cut in one or other of the cavity walls to prevent the cavity from forming a shorted loop. This approach presents difficulties when a sample tube stack is fitted, and another alternative is to accept the loss of the cavity body, but ensuring that it is at a minimum by making the window as large as possible. A window of this form is shown in Fig. 4.5(*b*). Here the microwave currents cut along xx and yy, but not along xy; this is thus something of a compromise. Where the window cuts current paths it is always necessary to place a soft metal gasket of silver, or some other suitable material, between the cavity and the window. Further reduction of the shorting effect of the cavity may be obtained by making the whole of both side walls windows. All edges of the windows then cut microwave current lines, so particular care has to be exercised in choosing a good gasket.

§4.5

SAMPLES OF SEVERE DIELECTRIC LOSS

When the E.S.R. sample has a significant dielectric loss, ϵ'' cannot be made zero in equations (4.21) and (4.22) and so the sensitivity is degraded. The deterioration can, however, be minimized by placing the sample carefully at one of the nodes of the electric field within the cavity, which, fortunately, invariably coincides with a region of maximum magnetic field. It then transpires from the following discussion that, generally, there is an optimum size of sample, above which the added dielectric loss resulting from the increased penetration of the sample into the electric field outweighs the gain due to the increase in sample volume. The problem of determining this optimum is somewhat complex in the general case, but is considerably simplified for those samples where the dielectric loss is so severe that the optimum thickness of sample is small compared with one wavelength. The most common example of this type of sample is an aqueous solution. In normal types of microwave cavity there are two distinct types of node in the electric field. In the rectangular cavity of Fig. 4.2(a) the node is clearly planar and demands a 'flat-cell' type of sample holder like that of Fig. 4.6. The cylindrical cavity of Fig. 4.2(b), on the other hand, has a node, which can be classed as 'linear', running down the axis of the cylinder. This type of cavity will therefore require a cylindrical sample

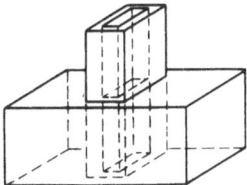

Fig. 4.6. Flat-cell aqueous sample holder for an H_{012} rectangular cavity.

holder. The two types of node are fundamentally different and are treated separately below. The effect of using a cylindrical sample in a rectangular cavity is also considered briefly.

Planar node

For the planar node, equations (4.21) and (4.22) give the sensitivity of the cavity for unsaturable and saturable samples respectively. Then, assuming the thickness of the sample to be small compared with one wavelength, it is permissible to assume that both H and dE/dx are independent of x, the distance from the node centre. With these assumptions, the quantities $\int_s H^2 dv$ and $\int_s E^2 dv$ in equations (4.21) and (4.22) can be written as $T_s \int_s H^2 da$ and $\int_s (dE/dx)^2 . da . \int_T x^2 dx$ respectively, where $\int_s da$ represents the integral over the area of the sample and $\int_T dx$ the integral over the thickness T_s of the sample. The latter term is then $T_s^3/12 . \int_s (dE/dx)^2 da$, so the cavity sensitivity equations become for the unsaturable sample

$$P_a^{1/2} = \frac{K_u f^2 T_s \int_s H^2 da}{\left(\dfrac{\pi\mu}{\sigma}\right)^{1/2} f^{1/2} \int_w H^2 da + \dfrac{2\pi f \epsilon_0 \epsilon'' T_s^3}{12} . \int_s \left(\dfrac{dE}{dx}\right)^2 da} P_i^{1/2} \quad (4.33)$$

and for the saturable sample

$$P_a^{1/2} = K_s f^2 \left[\frac{T_s^2 \int_s H^2 da \int_s h^2 da}{\left(\dfrac{\pi\mu}{\sigma}\right)^{1/2} f^{1/2} \int_w H^2 da + \dfrac{2\pi f \epsilon_0 \epsilon'' T_s^3}{12} \int_s \left(\dfrac{dE}{dx}\right)^2 da} \right]^{1/2} H_{sat} \quad (4.34)$$

Then, optimization with respect to sample thickness T_s gives for unsaturable samples

$$\left(\frac{\pi\mu}{\sigma}\right)^{\frac{1}{2}} f^{\frac{1}{2}} \int_w H^2 \, da \;=\; 2\frac{2\pi f \epsilon_0 \epsilon'' T_{opt}^3}{12} \int_s \left(\frac{dE}{dx}\right)^2 da \qquad (4.35)$$

and for saturable samples

$$2\left(\frac{\pi\mu}{\sigma}\right)^{\frac{1}{2}} f^{\frac{1}{2}} \int_w H^2 \, da \;=\; \frac{2\pi f \epsilon_0 \epsilon'' T_{opt}^3}{12} \int_s \left(\frac{dE}{dx}\right)^2 da \qquad (4.36)$$

Thus, for unsaturable samples the optimum sample thickness is that which gives a dielectric loss that is half the cavity loss, and for the saturable samples the thickness should be increased until the dielectric loss is twice the cavity loss. Experimentally these criteria can be judged by adding enough sample to reduce the Q-factor of the cavity to 2/3 of its original value in the first case, and to 1/3 of its original value in the second. It is therefore clear that if the same cavity is to be used for lossless samples, which may be either saturable or unsaturable, it is necessary to provide some means of varying the coupling from the cavity to the waveguide in order to provide a match in each case. Several methods of varying the coupling are used. One is to use a standard slide screw or three-screw tuner. These devices become somewhat lossy for anything but a small mis-match and really can only be satisfactorily used if the cavity is fitted with a number of different coupling irises. The iris which provides the nearest match is then used and the tuner is set to provide the final matching. A common method employed requires a thick iris with a dielectric screw that is inserted from the side of the coupling hole. Another method[8] is to place a short length of reduced height waveguide adjacent to the coupling hole on the outside of the cavity. The height of the waveguide is such that the waveguide is below cut-off when empty but above cut-off when filled with a dielectric plug. Then, as the dielectric is moved away from the hole the coupling is decreased. Ager, Cole and Lamb[9] have used a variable coupling probe and Faulkner and Holman[10] an adjustable waveguide matching stub outside the cavity.

To investigate the dependence of the sensitivity of a cavity with a planar node on the operating frequency and on the dimensions of the sample and cavity, it is necessary to insert the optimum sample thickness defined by equations (4.35) and (4.36) into the sensitivity relations of equations (4.33) and (4.34). This then gives for the unsaturable sample

$$
P_a^{1/2} = \frac{K_u f^2 \int_s H^2 da}{\frac{3}{2}\left[\left(\frac{\pi\mu}{\sigma}\right)^{1/2} f^{1/2}\int_w H^2 da\right]^{2/3}\left[\frac{2\pi f\epsilon_0 \epsilon''}{6}\int_s \left(\frac{dE}{dx}\right)^2 da\right]^{1/3}} P_i^{1/2}
\tag{4.37}
$$

and for the saturable sample

$$
P_a^{1/2} = K_s f^2 \left\{ \frac{\int_s H^2 da . \int_s \hat{h}^2 da}{3\left[\left(\frac{\pi\mu}{\sigma}\right)^{1/2} f^{1/2}\int_w H^2 da\right]^{1/3}\left[\frac{2\pi f\epsilon_0 \epsilon''}{24}\int_s \left(\frac{dE}{dx}\right)^2 da\right]^{2/3}} \right\}^{1/2} H_{sat}
\tag{4.38}
$$

Now the value of dE/dx is related to H by Maxwell's equations. When expressed in rectangular coordinates[1] Maxwell's third equation gives

$$
\frac{\partial E_y}{\partial x} - \frac{\partial E_x}{\partial y} = -\mu\frac{\partial H_z}{\partial t}
\tag{4.39}
$$

If the yz plane is taken to be the plane of the node and the z direction the direction of the H field, then, in the region of the node,

$$
\frac{\partial E_x}{\partial y} \simeq 0
$$

For oscillating fields

$$
\frac{\partial H}{\partial t} = 2\pi f H
\tag{4.40}
$$

and putting the above conditions into equation (4.39) gives

$$
2\pi f\mu H \simeq \frac{dE}{dx}
\tag{4.41}
$$

Then inserting equation (4.41) into equations (4.37) and (4.38) gives for unsaturable samples

$$
P_a^{1/2} \sim \sigma^{1/3} f^{2/3}\left(\frac{\int_s H^2 da}{\int_w H^2 da}\right)^{2/3} (\epsilon'')^{-1/3} P_i^{1/2}
\tag{4.42}
$$

and for saturable samples

$$P_a^{\frac{1}{2}} \sim \sigma^{1/12} f^{11/12} \left(\frac{\int_s H^2 da}{\int_w H^2 da} \right)^{1/6} \left(\int_s \hbar^2 da \right)^{\frac{1}{2}} (\epsilon'')^{-1/3} H_{sat} \qquad (4.33)$$

If the approximations are made so that

$$\int_s H^2 da = \frac{1}{2} H^2_{max} a_s$$

and

$$\int_w H^2 da = \frac{1}{2} H^2_{max} a_w$$

where a_s is the area of the sample and a_w the area of the cavity wall, then for unsaturable samples

$$P_a^{\frac{1}{2}} \sim \sigma^{1/3} f^{2/3} \left(\frac{a_s}{a_w} \right)^{2/3} (\epsilon'')^{-1/3} P_i^{\frac{1}{2}} \qquad (4.44)$$

and for saturable samples

$$P_a^{\frac{1}{2}} \sim \sigma^{1/12} f^{11/12} a_s^{2/3} a_w^{-1/6} (\epsilon'')^{-1/3} H_{sat} \qquad (4.45)$$

Thus in both cases the ratio of sample area to cavity area should be as high as possible. This is the only requirement regarding sample proportions for unsaturable samples, whilst, for saturable samples, the sample area should also be large absolutely. The dielectric loss ϵ'' varies with frequency in a manner which differs from sample to sample, and here it will only be possible to discuss the case of water. For water, Stoodley[11] gives

$$\epsilon'' \simeq 75 \left(\frac{f}{f_\epsilon} + \frac{f_\epsilon}{f} \right)^{-1} \qquad (4.46)$$

where $f_\epsilon = 16 \cdot 7$ Gc/s. Thus, for frequencies below 10 kMc/s, $P_a^{\frac{1}{2}} \sim f^{1/3}$ for unsaturable samples and $P_a^{\frac{1}{2}} \sim f^{7/12}$ for saturable samples, where the size of the magnet, and therefore the cavity and sample size, is fixed and $\epsilon'' \sim f$. Therefore, in both cases, f should be as high as possible. When the frequency is above, say, 20 kMc/s, $\epsilon'' \sim f^{-1}$ and the sample and cavity areas allowed by the magnet homogeneity are roughly proportional to f^{-2}. Then $P_a^{\frac{1}{2}} \sim f$ for unsaturable samples and $P_a^{\frac{1}{2}} \sim f^{1/4}$ for saturable samples, given that for both the ratio a_s/a_w remains constant.

Thus in all cases it would seem that the operating frequency should be as high as possible. Stoodley, however, calculated the optimum thickness of the sample cell for water to be 1/3 mm at 10 kMc/s. Clearly, the need to reduce the thickness for higher frequencies presents a problem. Then there is the noise figure of the detecting system which tends to deteriorate for frequencies above 10 kMc/s, so for unsaturable samples few attempts are made to improve the sensitivity of the cavity by raising the operating frequency. A question far more frequently asked is what deterioration there will be by going to much lower frequencies, by using Helmholtz coils in place of the normal magnet and by increasing the sample thickness. This procedure is, in particular, of interest when investigating biological samples or when using material that is too viscous or has to be sliced to get it into the sample cell. In these cases a cell thickness of 1/3 mm is clearly impracticable, and so the question of low-frequency operation will therefore be considered in more detail in §4.8.

Linear node

We shall now repeat the previous arguments in a form appropriate to a linear node. Here the sample is in the form of a cylinder of radius R_a and, for very lossy samples, it is permissible to assume that, within the sample, both H and dE/dr are independent of the distance r from the nodal line. With these assumptions

$$\int_s H^2 dv = \pi R_a^2 \int_L H^2 dl$$

and

$$\int_s E^2 dv = \int_L \left(\frac{dE}{dr}\right)^2 dl \cdot \int_A r^2 da$$

where $\int_L dl$ indicates the integral over the length of the sample and $\int_A da$ the integral over the cross-sectional area of the sample. Then, because

$$\int_A r^2 da = \pi R_a^4/2$$

equations (4.21) and (4.22) become for unsaturable samples

$$P_a^{1/2} = \frac{K_u f^2 \pi R_a^2 \int_L H^2 \, dl}{\left(\frac{\pi\mu}{\sigma}\right)^{1/2} f^{1/2} \int_w H^2 \, da + 2\pi f \epsilon_0 \epsilon'' \frac{\pi R_a^4}{2} \int_L \left(\frac{dE}{dr}\right)^2 \, dl} \, P_i^{1/2} \quad (4.47)$$

and for saturable samples

$$P_a^{1/2} = K_s f^2 \left[\frac{\pi^2 R_a^4 \int_L H^2 \, dl \cdot \int_L \mathcal{R}^2 \, dl}{\left(\frac{\pi\mu}{\sigma}\right)^{1/2} f^{1/2} \int_w H^2 \, da + 2\pi f \epsilon_0 \epsilon'' \frac{\pi R_a^4}{2} \int_L \left(\frac{dE}{dr}\right)^2 \, dl} \right]^{1/2} H_{sat} \quad (4.48)$$

Then optimization with respect to sample radius R gives for unsaturable samples

$$\left(\frac{\pi\mu}{\sigma}\right)^{1/2} f^{1/2} \int_w H^2 \, da = 2\pi f \epsilon_0 \epsilon'' \frac{\pi R_a^4}{2} \int_L \left(\frac{dE}{dr}\right)^2 \, dl \quad (4.49)$$

and for saturable samples R_a should be as large as possible. Thus, for unsaturable samples the optimum radius is that for which the dielectric losses are equal to the cavity losses, so that the cavity Q-factor is halved. For saturable samples the radius should be such that the dielectric losses are large compared with the cavity losses. Inserting the optimum radii into equations (4.47) and (4.48) gives for unsaturable samples

$$P_a^{1/2} = \frac{K_u f^2 \pi \int_L H^2 \, dl}{2 \left[\left(\frac{\pi\mu}{\sigma}\right)^{1/2} f^{1/2} \int_w H^2 \, da \cdot f \epsilon_0 \epsilon'' \pi^2 \int_L \left(\frac{dE}{dr}\right)^2 \, dl \right]^{1/2}} \, P_i^{1/2} \quad (4.50)$$

and for saturable samples

$$P_a^{1/2} = K_s f^2 \left[\frac{\pi \int_L H^2 \, dl \cdot \int_L \mathcal{R}^2 \, dl}{\pi f \epsilon_0 \epsilon'' \int \left(\frac{dE}{dr}\right)^2 \, dl} \right]^{1/2} H_{sat} \quad (4.51)$$

dE/dr is then related to H by Maxwell's third equation. When the equations are expressed in cylindrical coordinates, r, ϕ and z, one relation that emerges is

$$\frac{\partial E_\phi}{\partial r} + \frac{E_\phi}{r} - \frac{1}{r} \cdot \frac{\partial E_r}{\partial \phi} = -\mu \frac{\partial H_z}{\partial t} \quad (4.52)$$

In the region of the node $E_\phi = 0$, $\partial E_r / \partial \phi = 0$ owing to the cylindrical symmetry. Then for oscillating fields

$$\frac{\partial H}{\partial t} = 2\pi f H$$

so that, from equation (4.52),

$$2\pi f \mu H = \frac{dE}{dr} \tag{4.53}$$

Inserting equation (4.53) into equations (4.50) and (4.51) gives for unsaturable samples

$$P_a^{\frac{1}{2}} \sim \sigma^{\frac{1}{4}} f^{\frac{1}{4}} \left(\frac{\int_L H^2 dl}{\int_w H^2 da} \right)^{\frac{1}{2}} (\epsilon'')^{-\frac{1}{2}} P_i^{\frac{1}{2}} \tag{4.54}$$

and for saturable samples

$$P_a^{\frac{1}{2}} \sim f^{\frac{1}{2}} \left(\int_L R^2 dl \right)^{\frac{1}{2}} (\epsilon'')^{-\frac{1}{2}} H_{sat} \tag{4.55}$$

The approximations

$$\int_L H^2 dl \simeq \frac{1}{2} H_{max}^2 L_s$$

$$\int_w H^2 da \simeq \frac{1}{2} H_{max}^2 a_w$$

where L_s is the length of the sample and a_w the area of the cavity wall, can then be inserted to give for unsaturable samples

$$P_a^{\frac{1}{2}} \sim \sigma^{\frac{1}{4}} f^{\frac{1}{4}} \left(\frac{L_s}{a_w} \right)^{\frac{1}{2}} (\epsilon'')^{-\frac{1}{2}} P_i^{\frac{1}{2}} \tag{4.56}$$

and for saturable samples

$$P_a^{\frac{1}{2}} \sim f^{\frac{1}{2}} L_s^{\frac{1}{2}} (\epsilon'')^{\frac{1}{2}} H_{sat} \tag{4.57}$$

If we apply when the sample size varies with frequency similar criteria to those used for the planar node, we obtain for aqueous samples when the frequency is below 10 kMc/s $P_a^{\frac{1}{2}} \sim f^{-\frac{1}{4}}$ for unsaturable samples and $P_a^{\frac{1}{2}} \sim f^0$ for saturable samples, where the size of the cavity and sample is fixed and $\epsilon'' \sim f$. For frequencies above 20 kMc/s, we obtain $P_a^{\frac{1}{2}} \sim f^{5/4}$ for unsaturable samples and $P_a^{\frac{1}{2}} \sim f^{\frac{1}{2}}$ for saturable samples, where $\epsilon'' \sim f^{-1}$,

$L_s \sim f^{-1}$ and $a_w \sim f^{-2}$. It therefore appears that the normal operating frequency of 10 kMc/s is a poor choice for the linear node
and that varying the frequency in either direction should, in most
cases, give an improvement. However, the low-frequency improvement can hardly be realized, because, generally speaking,
a linear node requires a resonator that extends for at least a
quarter of a wavelength in all directions normal to the nodal
line. Such a resonator would be impossible to accommodate between the poles of a magnet at any frequency below 10 kMc/s.
The increase in sensitivity as the frequency is raised above
10 kMc/s is not particularly rapid and would doubtless be off-set
by a deterioration in the noise factor of the detecting system etc.
It is therefore unusual to attempt to vary the operational frequency from 10 kMc/s, as indeed it is unusual to use the linear type
of node at all. In fact there is little difference in sensitivity
between the H_{012} rectangular cavity and the H_{011} cylindrical
cavity when operated with optimum sample dimensions and
shapes. Stoodley[11] has made the necessary comparisons and
shows, for unsaturable samples, that, at 10 kMc/s, a planar
sample in an H_{012} rectangular cavity of uniform proportions
gives a voltage sensitivity of about 1·5 times that afforded by a
cylindrical sample in an H_{011} cylindrical cavity. The best that
can be obtained by putting the cylindrical sample in the rectangular cavity is about five times less than that obtained with the
planar sample. For aqueous samples the optimum radius of the
cylindrical sample in the cylindrical cavity is 0·4 mm and the
optimum thickness of the flat cell in the rectangular cavity is
0·3 mm.

§4.6

SMALL SAMPLES

The remaining class of sample to be discussed is the very small
sample, such as a single crystal of limited growth. Here the
sample size is fixed and will be assumed to be small compared
with the maximum size permitted by the magnet homogeneity considerations. It is then also reasonable to assume that the dielectric losses of the sample can be neglected. Equations (4.23)
and (4.24) are then valid for describing the sensitivity for

unsaturable and saturable samples respectively. Because the sample size is fixed, $\int_s \hbar^2 dv$ remains constant as the frequency is varied, whilst if the same type of resonator is used, $\int_w \hbar^2 da \sim f^{-2}$. Thus, $P_a^{1/2} \sim f^{7/2}$ for unsaturable samples and $P_a^{1/2} \sim f^{11/4}$ for saturable samples.

The relationship of $P_a^{1/2} \sim f^{7/2}$ for small fixed samples and the corresponding result of $P_a^{1/2} \sim f^{1/2}$ for unlimited samples [equation (4.28)] are well known, and were first obtained by Feher.[12] Usually the frequency dependence for the unlimited sample is considered not sufficiently rapid to overcome the deterioration in sensitivity due to the reduction in the noise factor of the detector, in the available klystron power and in the ratio of sample volume to sample holder volume, as the frequency is increased. For small fixed samples, however, the gain of $f^{7/2}$ outweighs these conditions and it is usual then to operate the spectrometer at the highest frequency possible.

The argument leading to the $f^{7/2}$ law is, however, based on the assumption that the cavity dimensions are all inversely proportional to frequency, or, more precisely, that the area of the cavity walls is proportional to $f^{-1/2}$. However, for a sample of a size small compared with one wavelength, it is possible to improve considerably the ratio of cavity wall area to sample volume by replacing the rectangular waveguide cavity with a coaxial cavity of a radius comparable with the sample size. The area of the resonator is then proportional to f^{-1}, so that $P_a^{1/2} \sim f^{5/2}$ for unsaturable samples and $P_a^{1/2} \sim f^{9/4}$ for saturable samples. The increase in sensitivity with frequency is then less dramatic but still probably worthwhile.

A measure which can further increase the sensitivity of the spectrometer for small samples is to load the coaxial line with a material of high dielectric constant ϵ' such as TiO_2 ($\epsilon' = 100$). The effect is to reduce the length of line by $(\epsilon')^{-1/2}$, so that for TiO_2 a further reduction of 10 should be obtained. Thus, by using these techniques, it is often possible to use a sample that is not too small and obtain much of the improvement that would otherwise only be obtained by going to higher frequencies. For excessively small samples however, it is invariably best to use the higher frequency.

§4.7

DIELECTRIC LOADING

The technique of loading the unused part of a spectrometer cavity can usefully be developed a little further. Generally speaking, the effect of introducing a dielectric into an electromagnetic field pattern is to reduce the wavelength within the dielectric by a factor $(\epsilon')^{-\frac{1}{2}}$, where ϵ' is the real part of the complex permittivity. The area of the cavity is therefore reduced by $(\epsilon')^{-1}$ for waveguide cavities and by $(\epsilon')^{-\frac{1}{2}}$ for thin coaxial cavities. For a given value of magnetic field at the sample, the cavity wall losses $R_s/2.\int_w H^2 da$ are therefore reduced accordingly and so, by equations (4.21) and (4.22), a gain in sensitivity is obtained. For this measure to be effective, it is, of course, important that the dielectric is not itself lossy. If ϵ'' is significant over the dielectric, the sensitivities given by equations (4.21) and (4.22) are reduced and the advantage is lost. Furthermore, it is very necessary to ensure that the dielectric is entirely diamagnetic and contains no paramagnetic impurities that will give a false E.S.R. signal. This point also applies to sample tubes, tail dewars, and any other component that is inserted into the cavity.

The method of dielectric loading is probably of little advantage for unlimited samples of zero dielectric loss, because the size of the sample and that of the cavity are then already comparable. The advantages of using small samples have been outlined and so all we have to consider is the case of samples with severe dielectric loss. For the rectangular cavity and planar sample cell of Fig. 4.6, the effect of loading the empty part of the cavity is to reduce the length by $(\epsilon')^{-\frac{1}{2}}$, as shown in Fig. 4.7(b). Thus the area of the side walls is reduced by $(\epsilon')^{-1}$ but the area of the end wall remains the same. The field value at the end wall is also the same, so that the effect of adding the dielectric is limited to reducing the side wall losses in the cavity. This limitation is largely overcome if the loading is restricted to only half a wavelength in the dielectric, as in Fig. 4.7(c). For a given value of H at the sample, the value of E at the faces of the dielectric is proportional to the impedance of the loaded waveguide, which is inversely proportional to

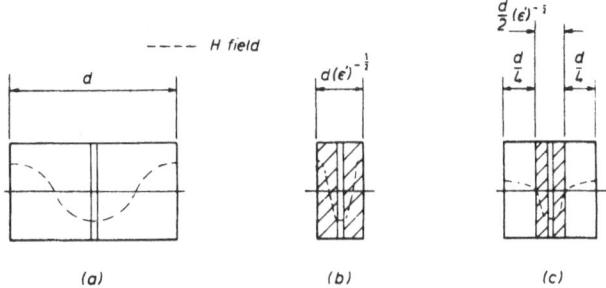

Fig. 4.7. Loading of an H_{012} rectangular cavity containing an aqueous sample cell with dielectric. (a) No loading, (b) fully loaded, and (c) loaded over central half wavelength.

$(\epsilon')^{\frac{1}{2}}$. Thus at the interface $E \sim (\epsilon')^{-\frac{1}{2}}$. Now in the fully loaded cavity the value of H at the end walls rises to the value at the sample, whilst, if empty end sections are used, the impedance is increased by a factor of $(\epsilon')^{\frac{1}{2}}$ over that of the loaded end sections, so the magnetic field rises only to a value of $H_{sample}(\epsilon')^{-\frac{1}{2}}$. Thus the end-wall losses, being proportional to H^2, are reduced by the factor $(\epsilon')^{-1}$. Furthermore, the losses in the side walls of the unloaded end sections are also reduced by the factor $(\epsilon')^{-1}$ over those of the completely unloaded cavity, since the magnetic field there is also reduced by $(\epsilon')^{\frac{1}{2}}$. Thus, over all sections of the cavity wall, both in the loaded centre section and the unloaded end sections, the wall losses are reduced by the factor $(\epsilon')^{-1}$. Consequently, if TiO_2 is used, an increase in voltage-signal-to-noise ratio of 100 should be obtained.

§4.8

LOW-FREQUENCY CAVITIES

As mentioned in §4.5, there are a few applications where there is some point in using an operating frequency well below the normal value of 10 kMc/s for an E.S.R. spectrometer. These are biological samples, which are aqueous and yet unsuitable for fitting in a 1/3-mm aqueous sample cell; *in-situ* electron-beam-irradiation experiments, where the electron beam must be directed along the axis of the magnetic field if it is not to be deflected; investigation of large samples, perhaps living organisms, which cannot be accommodated between the poles of a normal electro-magnet; problems where the E.S.R. data are required at the lower operating frequency; and training spectrometers where the con-struction is much simpler at the lower frequencies and the loss in sensitivity is of no real consequence.. For these cases it is therefore valuable to calculate how severe the loss of sensiti-vity will be when using a lower frequency. For samples of un-limited availability and zero dielectric loss, the comparison is obtained directly from equations (4.23) and (4.24), which are more easily interpreted if $\int_{s} \hbar^2 dv$ is put roughly equal to one half of the sample volume and $\int_{w} \hbar^2 da$ to one half of the area of the cavity wall.

It is, in practice, usually possible to design the cavity to fit the sample at almost all frequencies, by using either a loaded coaxial line at the higher frequencies or a simple tuned circuit at the lower frequencies. Equations (4.25) and (4.26) therefore apply and the reduction in sensitivity due to the $f^{3/2}$ term in the expression for unsaturable samples or to the $f^{7/4}$ term for satur-able samples is somewhat tempered by the increase in the sample size x_0. An increase in sample size is possible because the frequency is lowered and Helmholtz coils are used to pro-vide the magnetic field. For fixed samples this respite is un-fortunately not available and a very considerable loss is in-curred.

For aqueous or other lossy dielectric samples one or two arrangements have been used. The sample can be placed at the end of a coaxial cavity,[13] as in Fig. 4.8(c), or in a strip-line cavity, as in Fig. 4.8(a) and (b), or, at lower frequencies, in the coil of a tuned circuit.[14] Of these, the coaxial and strip-line cavities present planar nodes of electric field. Thus equations (4.44) and (4.45) are appropriate and for low frequencies where $\epsilon' \sim f$ give for unsaturable samples

$$P_a^{\frac{1}{2}} \sim f^{1/3} \left(\frac{a_s}{a_w} \right)^{2/3} P_i^{\frac{1}{2}} \tag{4.58}$$

and for saturable samples

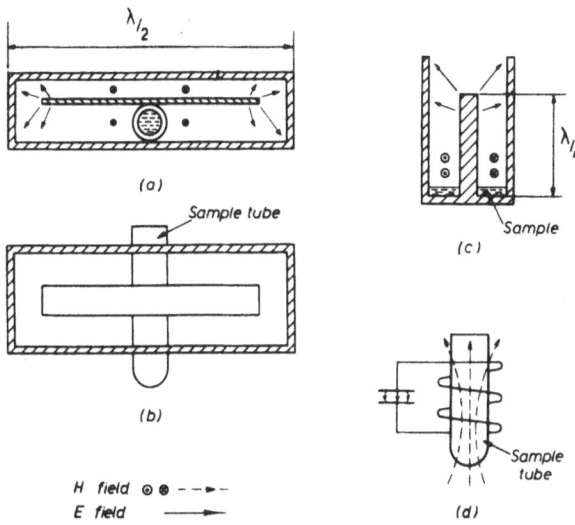

Fig. 4.8. Low-frequency spectrometer cavities: (a) section through strip-line cavity, (b) view of strip-line cavity with one wall removed (c) coaxial cavity and (d) tuned circuit.

$$P_a^{1/2} \sim f^{7/12} a_s^{2/3} a_w^{-1/6} H_{\text{sat}} \qquad (4.59)$$

Then, since a_s/a_w, the ratio of the sample area to the area of the cavity wall, is about the same at all frequencies, the reduction in sensitivity is proportional to $f^{1/3}$ for unsaturable samples and rather less than $f^{7/12}$ for saturable samples, because a_s can be increased as f is reduced. It is difficult to compare on the same basis the case of a simple coil since the coil has no node of electric field. Because there is an alternating potential across the ends of the coil, there must be an electric field, together with the magnetic field, along the axis of the coil. However, Kent and Mallard[14] in an experimental investigation obtained a sensitivity of $10^{-5}\Delta H$ mole per litre using 0·5 ml of aqueous sample at 100 Mc/s. Even when the somewhat lower operating power of 100 Mc/s is taken into account, the sensitivity they obtained compares unfavourably with the figure of $10^{-8}\Delta H$ mole per litre obtained on commercial spectrometers operating at 10 kMc/s and using a 0·3-mm flat cell. Calculations based on equations (4.33) and (4.34) reveal that, when using an oversized flat cell, the sensitivity deteriorates as the sample thickness increases (see Table 4.1). Thus for most biological samples it would seem better to operate at 10 kMc/s and to use an oversized sample cell than to lower the frequency.

Table 4.1 Deterioration of sensitivity as the thickness of a flat aqueous sample cell is increased from the optimum value of 0·3 mm.

	Sample thickness (mm)			
	0·3	0·65	1·40	3·0
Unsaturable samples	1	0·54	0·14	0·03
Saturable samples	1	0·66	0·32	0·15

§4.9

SLOW-WAVE STRUCTURES

Although the simple waveguide absorption cell used for gaseous microwave spectroscopy (see in §1.2) cannot be used for E.S.R. studies because the sample volume and sample holder required are extremely large, it is possible to reduce the volume to an acceptable figure and still retain most of the advantages of a travelling-wave system by substituting a slow-wave structure[15] for the spectrometer cavity. The simplest form of slow-wave structure is a normal length of waveguide operated almost at its cut-off frequency. It is well known[1] that one way of regarding the mode of propagation down a H_{01} rectangular waveguide is

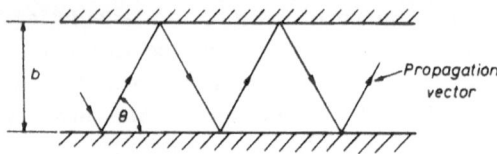

Fig. 4.9. Propagation along an H_{01} rectangular waveguide viewed as a series of reflected plane waves.

as a plane wave reflected successfully between the two narrow walls of the waveguide, as in Fig. 4.9. The angle θ that the propagation vector of the plane wave makes with the direction of propagation along the waveguide is given by

$$\cos\theta = \left[1 - \left(\frac{\lambda}{2b}\right)^2\right]^{\frac{1}{2}} \tag{4.60}$$

Thus, as λ, the free space wavelength, approaches $2b$, θ tends to $90°$ and the net rate of propagation down the waveguide approaches zero. The reduced velocity $c.\cos\theta$ is known as the 'group velocity' v_g and $\cos\theta$ as the 'slowing factor' S_s. Then, since energy is propagated along the waveguide more slowly, the stored energy per unit length is increased, for a given input power, by the factor S_s. Stored energy $\sim H^2$, so that $H^2 \sim S_s$. Then, by a slight modification of the arguments leading to equation (4.1) it can be shown that the power absorbed from the wave by the E.S.R. sample is equal to $\omega/2.\chi''\int_s H^2 dv$, so that the power absorbed per unit length by the sample is increased by S_s. From equation (4.10), the wall losses in the waveguide are proportional to H^2 so that the power absorbed per unit length by the walls of the waveguide is also increased by S_s. Thus the optimum length of waveguide given by optimizing equation (3.29) is reduced by the factor S_s, with no reduction in the overall power absorbed by the sample. It is shown in §3.5 that the sensitivity afforded by a travelling-wave sample cell of optimum length is very nearly equal to that of a resonant cavity so the slow-wave structure gives a sensitivity comparable with the resonant cavity. For the case of the near cut-off waveguide, the slowing factor can be raised to any desired figure by simply adjusting the klystron frequency to be sufficiently close to the cut-off frequency of the waveguide.

In practice, the most commonly used structure so far has been the helix, where the wave can crudely be considered to be slowed by following the turns of the helix and so obtaining a reduced axial velocity. The helix has the advantage of being an open wire structure and so presents no problems with regard to penetration of modulating field for the double-modulation method of detection. Furthermore, for E.N.D.O.R. studies, the helix can also be used as the coil for nuclear resonance. The disadvantage of the helix is that it has both electric and magnetic fields along the axis of the helix where the sample is usually placed.

Thus, with samples that have even moderate dielectric losses, the extra damping becomes very severe. There is little doubt that a better alternative would be the comb structure commonly used in travelling-wave masers and shown in Fig. 4.10. Here the magnetic field is separated from the electric field and thus the problems of sample dielectric loss are largely overcome. The main point of using a slow-wave structure is to overcome the problems associated with the frequency sensitivity of a resonant

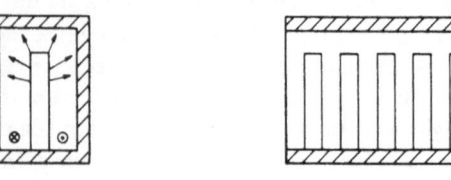

⊗ ⊙   *H field*
⟶ *E field*

Fig. 4.10. Sectional views of comb-type slow-wave structures.

cavity. However, it must not be assumed that slow-wave structures, or even waveguides, are entirely frequency-independent. To obtain a large slowing factor by simply adjusting the separation between the walls of a waveguide demands operation almost at cut-off. Thus a fairly small change in frequency would cause complete attenuation and, on the whole, attenuation would be fairly frequency-sensitive. Furthermore, the greater the slowing factor required the greater the frequency sensitivity is likely to be. However, it is certainly possible to devise structures, which for a size comparable with the equivalent resonant cavity, have a much reduced frequency sensitivity.

We have so far only discussed variation of attenuation with frequency. Ordinarily, a slow-wave structure would be operated as a transmission cell so that, for a given klystron power and an optimum structure attenuation of e^{-1}, the detector power would be fixed. Thus it is not possible to obtain the drive power to the detector for the optimum noise figure unless a bucking arm is introduced. If the structure is considered to be a microwave absorption cell of many wavelengths long, it should be clear that the change of phase ϕ between input and output for a given change in frequency will be very high. In fact

$$\frac{d\phi}{df} = \frac{2\pi N}{f}$$

where N is the number of electrical wavelengths of the cell. Consequently, the rapid change of phase with frequency of the wave through the structure, relative to the constant wave from the bridge arm, would probably produce as great a frequency sensitivity as the cavity would. Hence slow-wave structures are generally not a good substitute for a cavity, but can be used without too much disadvantage in applications such as the investigation of zero-field splittings (magnetic resonances between electrically split spin levels that require no magnet).

References

1. Ramo, S., and Whinnery, J.R., *Fields and Waves in Modern Radio* (Wiley, 1953).
2. Bjorken, J.D., and Bitter, F., *Rev. sci. Instrum.*, 1956, **27**, 1005.
3. Bennett, R.G., Hoell, P.C., and Schwenker, R.P., *Rev. sci. Instrum.*, 1958, **29**, 659.
4. Chester, P.F., Wagner, P.E., Castle, J.G., and Conn, G., *Rev. sci. Instrum.*, 1959, **30**, 1127.
5. Firth, I.M., *J. sci. Instrum.*, 1962, **39**, 131.
6. Lamb, J., and Ager, R., *Rev. sci. Instrum.*, 1959, **30**, 559.
7. Cook, A.R., Mataresse, L.M., and Wells, J.S., *Rev. sci. Instrum.*, 1963, **35**, 114.
8. Gordon, J.P., *Rev. sci. Instrum.*, 1961, **32**, 658.
9. Ager, R., Cole, T., and Lamb, J., *Rev. sci. Instrum.*, 1963, **34**. 308.
10. Faulkner, E.A., and Holman, A., *J. sci. Instrum.*, 1963, **40**, 205.
11. Stoodley, L.G., *J. Electr. Cont.*, 1962, **14**, 531.
12. Feher, G., *Bell Syst. Tech. J.*, 1957, **36**, 449.
13. Cook, R.F., and Stoodley, L.G., *Inst. J. of Radiation Biology*, 1963, **7**, 155.
14. Kent, M., and Mallard, J.R., *J. sci. Instrum.*, 1965, **42**, 505.
15. Siegman, A.E., *Microwave Solid-State Masers* (McGraw Hill, 1964).

Chapter 5

Superheterodyne Spectrometers

It was shown in §3.14 that the upper limit to the frequency at which the magnetic field can be modulated in a double-modulation spectrometer is determined by the linewidth of the magnetic resonance. For the majority of samples a value of 100 kc/s is permissible and, as we concluded in §2.7, does not involve too serious a deterioration in the noise figure of the detecting system compared with the noise figure that could be obtained if the choice of modulating frequency were unlimited. For very narrow lines, however, it is necessary to reduce the modulating frequency further, but then the deterioration in the noise figure of a conventional double-modulation system becomes severe. Another condition often requiring the modulating frequency to be reduced from 100 kc/s is that of operating at very low temperatures ($\sim 4°K$). The conductivity of the cavity walls is then increased and, if external modulation coils are used, it becomes difficult to obtain adequate penetration of the modulating field in the cavity. In these situations superheterodyne detection allows the modulating frequency to be reduced without incurring an increase in the noise figure.

§5.1

BASIC SUPERHETERODYNE SPECTROMETER

The basic method of applying superheterodyne detection to a spectrometer is shown in Fig. 5.1. Here the system is a conventional reflection-cavity spectrometer without microwave bucking, but a portion of the output from a second klystron, together with the output from the cavity and circulator, is fed to the microwave detector D_1. The two signals mix in the detector and produce at the output a component having a frequency equal to the difference

142

between the frequencies of the two klystrons. D_1 thus becomes a mixer and the second klystron acts as a local oscillator for the signal originating from the first klystron. (The first klystron will

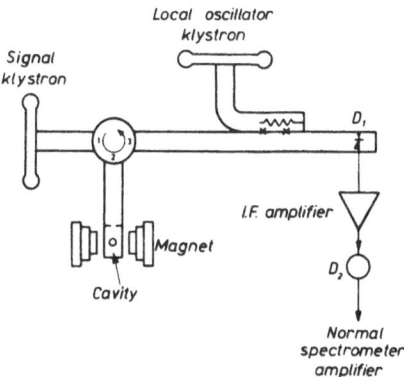

Fig. 5.1. E.S.R. spectrometer with superheterodyne detection.

hereafter be referred to as the signal klystron and the second as the local oscillator klystron.) The a.c. signal at the output of the microwave detector is then passed to an a.c. amplifier suitably tuned to the difference or intermediate frequency (i.f.) between the two klystrons, and is finally rectified by a second detector D_2.

The reason for employing superheterodyne detection is that the amplifier following the microwave detector can be arranged to have almost any desired frequency by adjusting the frequency separation between the two klystrons. Thus, because the frequency of the amplifier following the microwave detector is independent of the field-modulation frequency, its value can be chosen solely to give the best noise factor. In practice, the value of intermediate frequency giving the best noise factor is in the region of 30 Mc/s.* From Fig. 2.24(a), it is clear that this value generally avoids the effects of flicker noise from the microwave detector, and is not sufficiently high to cause diffi-

---

* Recent advances in crystal technology suggest that lower values can now be used without disadvantage.

culties in obtaining a good noise figure for the i.f. amplifier.

In spite of the practical arrangements, it is helpful to regard the superheterodyne spectrometer as though the detector from a normal spectrometer has been removed, the local oscillator, mixer and i.f. amplifier added as new components, and the original detector moved to the position of D_2. It is then clear that the function of the local oscillator and mixer is simply that of a frequency changer which converts the microwave output from the spectrometer to a frequency suitable for amplification. The function of D_2 is then seen to be no different from that of the detector of a conventional system, and the remainder of the spectrometer is therefore exactly the same as that which follows the microwave detector of the equivalent conventional system. It is also clear that the noise factor of the superheterodyne detecting system is just that of a conventional superheterodyne microwave receiver, whose noise factor is given in Fig. 2.24(a).

Because the second detector of a superheterodyne spectrometer performs the same function as the microwave detector of a conventional spectrometer, it is necessary to have a steady a.c. component that reaches the second detector of the superheterodyne system and corresponds to the bucking signal of the conventional system. It is shown in §§2.6 and 2.7 that a bucking signal is normally required to ensure an adequate efficiency of detection and also to allow discrimination between the absorptive and dispersive parts of the magnetic resonance signal. The fact that the frequency is changed before detection makes no difference to these considerations, and the bucking signal can, in principle, be provided by any of the methods used in a conventional spectrometer. However, the gain of the i.f. amplifier in the superheterodyne system usually needs to be fairly high, and, therefore, the level of the microwave bucking signal needed to provide a suitable amplitude at the input of the second detector is much lower than that required to drive the microwave detector of a conventional system. It is therefore practical to dispense with the normal microwave bucking arm in the conventional system and to provide the bucking by slightly mis-matching or mis-tuning the spectrometer cavity. The degree of mis-adjustment of the cavity required with superheterodyne detection is far too small to influence the sensitivity of the cavity.

§5.2

DISADVANTAGES OF THE SUPERHETERODYNE

The main disadvantage of the superheterodyne spectrometer of Fig. 5.1 is that the amount by which the signal klystron can be allowed to drift from the resonant frequency of the sample cavity is much smaller than for a conventional system. In Fig. 5.2, OA represents the bucking signal reaching the detector of a conventional spectrometer and AB($\simeq$ AC) the reflection from the cavity due to a degree of mis-tuning. AB then combines with OA at the detector to provide a resultant signal OB. The effect of f.m.

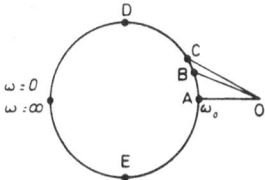

Fig. 5.2. Effect of the bucking level on the conversion of klystron f.m. noise to a.m. noise.

noise modulation of the klystron makes the resultant fluctuate between the limits of OB and OC. To ensure that this fluctuation does not result in a corresponding fluctuation in the amplitude of the resultant, it is important to keep the degree of mis-tuning sufficiently small to ensure that $|AB| \ll |OA|$. For a superheterodyne system, the amplitude of the bucking signal is much reduced, so a correspondingly smaller degree of mis-tuning can be tolerated. The same argument can be extended to the suppression of the dispersive part of the magnetic resonance signal, because this can also be represented as a change in the reflected signal from the cavity, i.e. a change from AB to AC.

In some ways the increased sensitivity to de-tuning may be considered as a reduction in the effective bandwidth of the cavity. From equation (3.8) the reflection coefficient ρ_M of a reflection cavity which is matched at the resonance frequency ω_0 is, for other frequencies ω, given by

$$\rho_M = \frac{jX}{2R + jX}$$

where R is the effective resistance of the cavity and X its reactance. Then, for resonators having a high unloaded Q-factor Q_U, $X \simeq 2Q_U R\, \delta\omega/\omega_0$ if $\delta\omega = \omega - \omega_0$. Thus

$$\rho_M = \frac{jQ_U \dfrac{\delta\omega}{\omega_0}}{1 + jQ_U \dfrac{\delta\omega}{\omega_0}}$$

Then, if ρ_B representing the bucking signal is added to ρ_M, the combined signal reaching the microwave mixer or detector is proportional to a combined transfer function ρ_D, where

$$\rho_D = \rho_B + \frac{jQ_U \dfrac{\delta\omega}{\omega_0}}{1 + jQ_U \dfrac{\delta\omega}{\omega_0}}$$

Then, provided that $\rho_B \ll 1$, which it certainly is in the superheterodyne, the value of $\delta\omega$ at which $|\rho_D|$ rises from its value ρ_{DO} at resonance to a value of $\sqrt{2}\rho_{DO}$ is such that the effective bandwidth $(\Delta\omega)_{\text{eff}}$, which is usually taken as twice the above value of $\delta\omega$, is given by

$$(\Delta\omega)_{\text{eff}} = 2\omega_0 \frac{\rho_{DO}}{Q_U} \tag{5.1}$$

In contrast, the frequency separation $(\Delta\omega)_L$ between points D and E of Fig. 5.2, which is the bandwidth of the resonator as normally defined, is given by the difference between the two values of $\delta\omega$ for which $Q_U \delta\omega/\omega_0 = 1$. Thus

$$(\Delta\omega)_L = \frac{2\omega_0}{Q_U} \tag{5.2}$$

The effective bandwidth of the cavity is therefore the normal loaded bandwidth $2\omega_0/Q_U$ divided by the attenuation of the bucking arm. Thus, the greater the attenuation required the sharper the effective bandwidth.

The above reduction in bandwidth is not entirely absent in conventional reflection-cavity systems. Fig. 2.24(a) shows that to obtain the optimum noise factor for the detecting system of a conventional spectrometer, a microwave drive of $0·5$ mW is required. Klystron powers, however, are generally about 50 mW, so that a ρ_{DO} value of about 1/10 is normally required. Thus for a cavity having a natural bandwidth of say 3 Mc/s, the drift needs to be small compared with 300 kc/s, rather than with β Mc/s.

In the superheterodyne system the required value of ρ_{DO} depends on the i.f. amplifier gain, which ideally should be as low as possible. However, there must be sufficient gain to overcome the noise originating from the second detector and subsequent amplifier.

A well-known formula which relates the overall noise figure F of a system to the noise factors, F_1, F_2 ... F_n, of its component parts is

$$F = F_1 + \frac{F_2 - 1}{G_1} + \frac{F_3 - 1}{G_1 G_2} ... + \frac{F_n - 1}{G_1 ... G_{n-1}} \quad (5.3)$$

where G_r is the available power gain of the rth stage. If the complete detecting system of a superheterodyne spectrometer is divided into two sections at the input of the second detector, then F_1 represents the noise factor of the frequency changer and the i.f. amplifier, and F_2 the noise factor of the second detector and subsequent amplifier. Thus F_2 is approximately equal to the noise factor of a conventional spectrometer which uses the same value of field-modulation frequency as the superheterodyne. From equation (5.3) it is therefore clear that, if the overall noise factor is not to be degraded below the value of F_1, G_1 the gain of the frequency changer and amplifier must be large compared with the ratio of $(F_2 - 1)$ to F_1. In fact, F_2 is usually much greater than unity so that G_1 has, in effect, to be large compared with the ratio of F_2 to F_1. To keep the required value of G_1 low, it is therefore necessary that F_2 should not be unduly high, and it is thus important to keep the field-modulation frequency f_m as high as possible.

The extent to which superheterodyne detection can increase the problems of frequency drift in a spectrometer can be illustrated by considering the example of a double-modulation spectrometer with a field-modulation frequency of 100 c/s. This value is generally about the highest that will allow the modulating field to penetrate the walls of a microwave cavity of conventional wall thickness. The ratio of the noise figure of a conventional spectrometer with a modulating frequency of 100 c/s is, from Fig. 2.24(a), about 40 dB higher than that of a 30 Mc/s superheterodyne. Thus G_1 needs to be 40 dB plus, say, a 10 dB safety margin, and the attenuation ρ_{DO} of the microwave bucking signal therefore has to be 50 dB more than for the conventional system. Thus the degree of mis-tuning permitted is reduced from 300 kc/s, deduced above for the conventional system, to a value of at least 1 kc/s. It is therefore clear that, in general, some kind of automatic frequency control (A.F.C.) system, which ensures that the main klystron is exactly tuned to the resonant frequency of the cavity, is essential for a superheterodyne spectrometer. Even with this provision, however, the basic superheterodyne system is notorious for being susceptible to frequency drift, microphonics, etc., and experience has led to its avoidance wherever possible.

It must, nevertheless, be allowed that, both when working at very low temperatures and when linewidths are very small, samples tend to saturate at a lower power. Under these conditions the problems are less.

§5.3

DRIVE TO THE SECOND DETECTOR

In the above discussion it has been assumed that the level of drive to the second detector of a superheterodyne spectrometer should be that which gives the optimum combined noise figure for the second detector and subsequent amplifier. This is not really the case. In reality we wish to make ρ_{DO} as high as possible in order to minimize the effects of frequency drift, whilst, at the same time, the noise from the i.f. amplifier must override that from the second detector and subsequent amplifier. In the following argument the level of drive allowing the maximum value of ρ_{DO} is determined and found not to be critical for drive levels above the value of the optimum noise factor of the second detector.

If N_0 represents that part of the noise power at the output of the amplifier following the second detector which originates from the amplifier and second detector, and if N_{in} represents the

noise from the microwave mixer and intermediate amplifier referred to the input of the microwave mixer, then the requirement
expressed above is that N_{in} multiplied by the overall gain of the
system should be, say, ten times N_0. The overall gain may be
written as $G_m G_d G_a$, where G_m is the power gain of the microwave mixer and i.f. amplifier, G_d the power gain of the second
detector, and G_a the power gain of the amplifier which follows
the second detector. Thus

$$N_{in} G_m G_d G_a = 10 N_0 \qquad (5.4)$$

Of these factors, both N_0 and G_d are functions of the drive
power P_i at the input to the second detector. Also $P_i \sim \rho_{DO}^2 G_m$,
so that, from (5.4),

$$\rho_{DO}^2 \sim \frac{P_i N_{in} G_a G_d (P_i)}{N_0 (P_i)} \qquad (5.5)$$

F_2, the noise factor of the second detector and subsequent
amplifier, is given by

$$F_2 = \frac{N_0}{kTBG_d G_a} \qquad (5.6)$$

Combining equations (5.5) and (5.6) and expressing the result in
decibels gives

$$\rho_{DO}(\mathrm{dB}) = A + P_i(\mathrm{dB}) - F_2(\mathrm{dB}) \qquad (5.7)$$

where A is a constant. Thus ρ_{DO} is a maximum at the value of
P_i for which $dF_2(\mathrm{dB})/dP_i(\mathrm{dB})$ is equal to unity, i.e. when the
slope of the curve corresponding to Fig. 2.24(a) for the second
detector is equal to unity. Fig. 2.24(a) then shows that the slope
of the curve is roughly equal to unity for all measured values of
drive above the value for the optimum noise figure. Over this region
the value of ρ_{DO} therefore remains constant.

A method of obtaining the correct value of ρ_{DO} is first to make
the gain of the i.f. amplifier very low, and then to increase ρ_{DO}
until the drive to the second detector is somewhat above that
which gives the optimum noise figure for the second detector and
subsequent amplifier. The level of drive which gives the optimum
noise figure for the second detector and following amplifier is
easily determined by disconnecting the input of the second de-

tector from the output of the i.f. amplifier, and by coupling the second detector to an oscillator tuned to the intermediate frequency. Fig. 2.24 then shows that the level of drive for the optimum noise figure corresponds roughly to the level at which the noise observed at the output of the amplifier following the second detector begins to increase. By varying the output level from the oscillator, this point can be determined. The reason for replacing the signal from the output of the i.f. amplifier by that from an oscillator, is to avoid the added contribution of noise from the i.f. amplifier, which would otherwise confuse the measurement. Then, as the value of ρ_{DO} can be adjusted to give the correct level of drive to the second detector, that is, a level which is somewhat above that which gives the optimum noise figure for the second detector and subsequent amplifier, the next step is to increase the gain of the intermediate amplifier and, at the same time, to reduce ρ_{DO} without altering the drive to the second detector. The effect of this method of increasing the gain of the i.f. amplifier is to increase the contribution of noise from the microwave mixer and i.f. amplifier relative to that from the second detector and following amplifier, whilst, at the same time, keeping the level of drive at the somewhat noncritical value which ensures that the value of ρ_{DO} is as high as possible. The process is therefore continued until the noise from the i.f. amplifier and microwave mixer rises above that from the second detector and following amplifier, and exceeds it by the safety factor of 10 dB. Thereafter, the gain of the i.f. amplifier should be unaltered and the correct value of ρ_{DO} can always be obtained by adjusting the matching of the spectrometer cavity to give the same level of drive to the second detector.

§5.4

KLYSTRON AMPLITUDE-MODULATED NOISE

In §§3.9 and 3.10 it is shown that klystron noise can add to the overall noise level from a conventional spectrometer in three ways. Firstly, the a.m. noise from the klystron can contribute directly. Secondly, it is possible for the microwave circuitry to convert the f.m. noise from the klystron into a.m. noise. Finally, the a.m. noise can in some circumstances be 'enhanced' by the microwave system. The question of how these types of noise add to the noise level in a superheterodyne spectrometer must therefore be considered.

In the conventional reflection-cavity spectrometer the a.m. noise sidebands from the klystron are attenuated by the factor ρ_{DO}. It is shown in §2.7 that, if ρ_{DO} is adjusted to give the optimum level of microwave drive to the detector, the contribution of the normal klystron a.m. noise is generally a little too low to be of significance. In the superheterodyne spectrometer this situation is fortunately, if anything, improved. The effect of converting from a conventional spectrometer to a superheterodyne spectrometer is usually to improve the noise factor of the detecting system by some factor a_i. Thus the klystron noise might be expected to figure more prominently. However, it is shown in §5.2 that, on changing to superheterodyne detection, the value of ρ_{DO} must be decreased to give an added power loss of rather more than a_i. The a.m. noise sidebands are subject to this reduction and thus the klystron a.m. noise is reduced even more than the detector noise.

§5.5

KLYSTRON FREQUENCY-MODULATED NOISE

The tendency for a superheterodyne spectrometer to convert f.m. noise from the klystron to a.m. noise is much greater than it is for a conventional spectrometer. If ρ_{DO} is taken to represent OA in Fig. 5.2, ρ_x to represent the added reflection AB from the cavity due to a degree of mis-tuning, $\delta\rho_x$ the fluctuation in ρ_x due to the klystron f.m. noise, and ρ the resultant reaching the detector, then if $\rho_x \ll \rho_{DO}$ the resulting fluctuation $\delta|\rho|$ in the amplitude of ρ is given by

$$\delta|\rho| \simeq \frac{|\rho_x|}{|\rho_{DO}|}|\delta\rho_x| \tag{5.8}$$

Thus the efficiency of conversion is inversely proportional to ρ_{DO}; so, unlike the a.m. noise, the reduction of which is greater than the improvement in the noise factor a_i when superheterodyne detection is used, the converted f.m. noise power is *increased* by rather more than a_i. Thus, to obtain the full potential improvement of the superheterodyne system, it is again seen that the signal klystron must be more carefully tuned to the resonant frequency of the cavity.

The question of converting a conventional spectrometer with a field-modulation frequency of 100 c/s to superheterodyne detection again provides an example to illustrate the magnitude of the above problem. In this instance a_i = 40 dB or a voltage ratio of 100. Thus ρ_{DO} must be reduced by about 300 and then, by equation (5.8), the degree of off-tuning, represented by ρ_x, which will give the same efficiency of conversion, is also reduced by 300. However, it is of little use merely to leave the level of klystron f.m. noise unchanged. The level must, in fact, be further decreased by the factor a_i to bring it below the reduced level of the detector noise afforded by the superheterodyne system. Thus, the degree of frequency drift allowed is, in reality, reduced by a factor greater than 30 000. It is therefore not surprising that the potential sensitivity of the basic superheterodyne system is seldom realized.

§5.6

ENHANCED AMPLITUDE-MODULATED NOISE

Since the effect of enhanced a.m. noise is shown in §3.10 to become more severe as the bandwidth of the spectrometer cavity is reduced, it is conceivable that the reduced effective bandwidth of the cavity in a superheterodyne spectrometer might increase the enhancement beyond that obtained in a conventional spectrometer. This, however, is fortunately not the case. Taking the example of a conventional spectrometer with a field-modulating frequency of 100 kc/s, it is found in practice that enhanced a.m. noise does, just noticeably, increase the overall noise level. If then the modulating frequency is reduced to, say, 100 c/s, and a superheterodyne with an intermediate frequency of 100 kc/s is used to maintain the original noise figure, then the power gain of the microwave mixer and amplifier needs to be roughly ten times the ratio of the noise factors at the two frequencies. From Fig. 2.24(*a*) this ratio is, in fact, the inverse of the ratio of the two frequencies. Thus the rejection of the cavity system ρ_{DO} has to be increased by the factor $[100 \text{ c/s}/(10 \times 100 \text{ kc/s})]^{1/2}$, and the effective bandwidth of the cavity is reduced accordingly. In contrast, the separation of the noise sidebands from the carrier is reduced in direct proportion to the two frequencies, that is, by a factor of 1000, so that the noise sidebands close in towards the carrier frequency more rapidly than the cavity rejection curve narrows. The enhancing of the a.m. noise sidebands is therefore less severe in a spectrometer with superheterodyne detection than it is when conventional detection is used. The figure of 100 kc/s for the intermediate frequency of the superheterodyne system was chosen to simplify the argument. In practice, a value of 30 Mc/s is more probable, but the further reduction of ρ_{DO} then required, which is equal to the inverse of the improvement in noise factor between 100 kc/s and 30 Mc/s, is still not enough to cause the klystron noise to be significantly enhanced.

§5.7

LOCAL OSCILLATOR NOISE

In the superheterodyne detecting system of a spectrometer, as in any microwave superheterodyne, it is possible that the local oscillator will generate noise sidebands of sufficient amplitude so that those separated from the frequency of the local oscillator by the intermediate frequency can mix with the main output from the local oscillator in the microwave mixer and produce at the intermediate frequency noise components large enough to increase the overall noise level. The use of a balanced mixer in the manner shown in Fig. 5.3 overcomes the problem quite adequately. Since the two detectors in Fig. 5.3 have opposite polarity, the combined d.c. output from the detectors, which is due to the local oscillator, is zero, and fluctuations in the amplitude of the local oscillator are therefore of no effect. The characteristics of the hybrid junction, however, ensure that either the path between the local oscillator and the two detectors, or the path between the spectrometer cavity and the two detectors, has an effective difference of one half wavelength. Thus, when the microwave signal from the spectrometer cavity beats with that from the local oscillator to produce a periodic variation in the amplitude of the combined

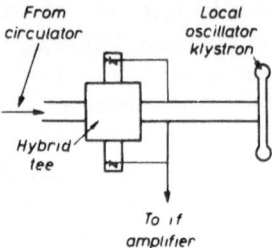

Fig. 5.3. Balanced mixer for a superheterodyne spectrometer.

signal at the intermediate frequency, the phase of the beat is opposite for the composite signals reaching the two detectors. Then, again because of the different polarity of the two detectors, the resulting components at the intermediate frequency at the outputs of the two detectors are in phase and add in the i.f. amplifier.

Apart from the question of local oscillator noise, there is no difference between the noise factor of a balanced mixer and a single-ended mixer. If Fig. 5.4(a) represents the output circuit of a single-ended mixer, where E_s is the signal voltage, $\overline{e_n^2}$ the noise voltage and R_0 the output impedance, then Fig. 5.4(b)

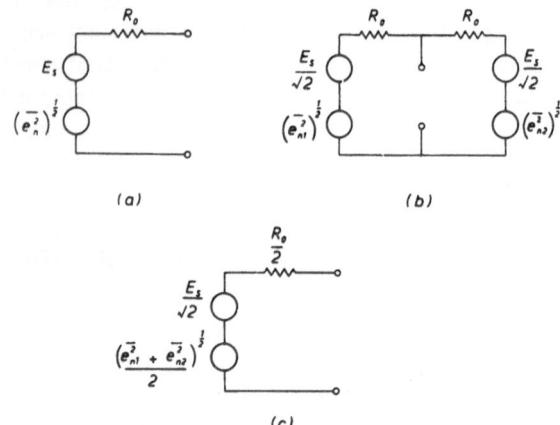

Fig. 5.4. Equivalent circuits of mixer outputs for (a) a single-ended mixer, and (b) a balanced mixer. (c) An equivalent circuit for the transform of (b).

represents the output from the equivalent balanced mixer. E_s is divided by $\sqrt{2}$ in (*b*) because of the division of signal power at the hybrid junction. The Thevenin transform of Fig. 5.4(*b*) is then as shown in Fig. 5.4(*c*) and indicates that, if $\overline{e_{n1}^2} = \overline{e_{n2}^2}$, there is no difference between the signal-to-noise ratio at the output of (*a*) and (*c*). Moreover, the conversion gain is the same because the available signal output power is equal to $|E_s^2|/8R_0$ in each case.

§5.8

IMAGE NOISE

Another way in which the klystron noise can add to the overall noise level of a superheterodyne spectrometer is via the image channel, as shown in Fig. 5.5. Here, Fig. 5.5(*a*) shows the carrier from the signal klystron, together with the associated noise sidebands. Fig. 5.5(*c*) shows the local oscillator signal, which is separated in frequency from the signal klystron by the intermediate frequency. The microwave superheterodyne then accepts noise over two frequency ranges centred at frequencies which differ from the local oscillator frequency by the intermediate frequency, and which have widths equal to the bandwidth of the i.f. amplifier. One of these regions of acceptance is centred at the

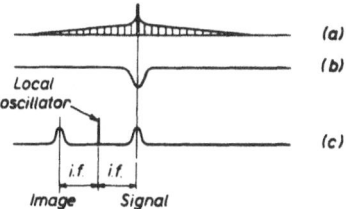

Fig. 5.5. Mechanism whereby signal klystron noise can contribute via the image response of the superheterodyne detector. (*a*) Signal klystron and noise sidebands, (*b*) rejection of spectrometer bridge, and (*c*) local oscillator and total input response of superheterodyne.

frequency of the signal klystron and is termed the 'signal' channel, whilst the other is termed the 'image' channel. Both channels are shown in Fig. 5.5(*c*). The noise sidebands from the signal klystron at the image frequency are very much weaker than those at the signal frequency, but the signal frequency sidebands are subject to the rejection of the cavity shown in Fig. 5.5(b), whilst the image sidebands are not. Thus, the image noise can represent a significant contribution to the overall noise level.

§5.9

FREQUENCY STABILIZATION OF THE LOCAL OSCILLATOR

Generally the bandwidth of the i.f. amplifier in a superheterodyne spectrometer can be made a few megacycles without difficulty. Consequently, the problem of maintaining the correct separation between the local oscillator klystron and the signal klystron presents little difficulty. It is sometimes thought desirable to stabilize the frequency of the local oscillator klystron relative to that of the signal klystron, but this is seldom really necessary.

The factor that limits the bandwidth of the i.f. amplifier is its noise factor. Bandwidths of above 20 per cent of the operating frequency present some difficulty when devising a suitable network to match the mixer to the input of the amplifier, and a deterioration can result. Apart from this consideration, widening the bandwidth does not add to the overall noise at the output of the spectrometer. As the second detector can also be regarded as a mixer, with the converted component from the signal klystron acting as the local oscillator, the effective noise input response to the second detector can be represented by a diagram similar to Fig. 5.5(c). The two regions of band-pass then have the bandwidth of the tuned amplifier following the second detector, which is tuned to the field-modulating frequency f_m. The separation of the two channels is therefore equal to $2f_m$ and, provided that the bandwidth of the intermediate amplifier is large compared with this figure, any further increase simply presents to the second detector additional noise outside its range of acceptance.

When stabilization of the frequency of the local oscillator is required, the normal method used for microwave receivers is quite satisfactory. This method is shown in Fig. 5.6. Here a small portion of the output from the i.f. amplifier is further amplified and fed to a discriminator. The output from the discriminator then corrects any drift in the frequency of the local oscillator klystron by controlling its reflector voltage. Because the output from the i.f. amplifier is rather strongly dependent on changes

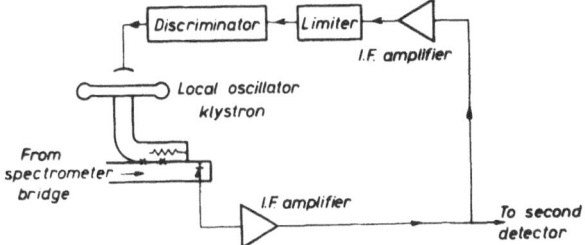

Fig. 5.6. A.F.C. system for local oscillator of superheterodyne
spectrometer.

in the frequency of the signal klystron relative to the cavity, it
is also necessary to have an effective amplitude limiter before
the discriminator. Another way of avoiding the amplitude changes
is to mix a portion of the output from the local oscillator klystron
with a small portion of the output from the signal klystron taken
prior to the cavity, as in Fig. 5.7.

§5.10

THE SUPERHETERODYNE SYSTEM OF
HIRSHON AND FRAENKEL

Hirshon and Fraenkel[2] have extended the system of Fig. 5.7 so
that not only can the frequency of the local oscillator be stabilized at the correct separation from that of the signal klystron,
but also the frequency of the signal klystron can be stabilized
relative to the spectrometer cavity. When the local oscillator
stabilizer of Fig. 5.7 is working, and the signal klystron frequency changes, the phase and amplitude of the i.f. signal at the
output of the A.F.C. mixer is not much altered. In contrast, the
signal at the output of the signal mixer varies very rapidly with
frequency, because it retains all the phase and amplitude variations to which the reflected signal from the spectrometer cavity
is subject, when the frequency of the signal klystron varies.
These variations are those summarized in Fig. 5.2, and thus it is
clear that if the output from the A.F.C. amplifier is used as a
reference for a phase-sensitive detector, to which the output of
the signal amplifier is made the input, then, by suitably adjusting the phase of either reference or signal, the phase-sensitive
detector can be made to respond to any one phase component of

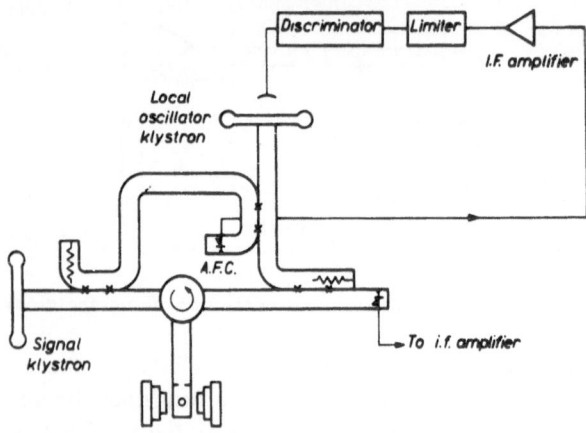

Fig. 5.7. Improved A.F.C. system for local osoillator.

Fig. 5.2. Then, if the phase relationship is adjusted to accept the components in quadrature with the phase of OA, a signal proportional to the degree of mis-tuning is developed at the output of the phase-sensitive detector. This d.c. output can then be amplified and used to control the reflector voltage of the signal klystron, and thus to ensure that the quadrature or reactive component of the reflection coefficient from the cavity is zero, i.e. that the cavity is 'on tune'.

In principle, the phase adjustment could be made in either i.f.

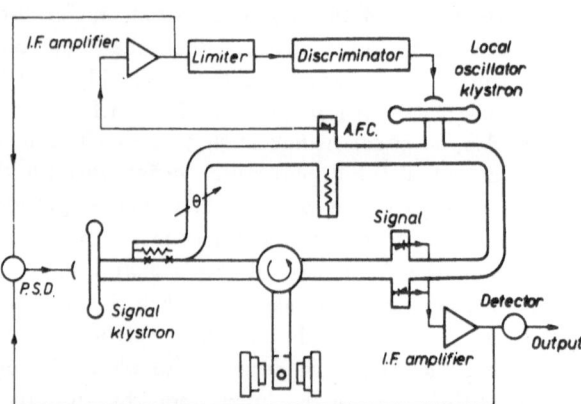

Fig. 5.8.   Superheterodyne   spectrometer   incorporating   the Hirshon–Fraenkel system of frequency stabilization.

amplifier but, since microwave phase shifters are somewhat easier to design than those for radio frequencies, the same effect can more conveniently be obtained by shifting the phase of the microwave signal from the signal klystron. A system following the broad lines of that of Hirshon and Fraenkel is shown in Fig. 5.8. In common with the original, this uses a balanced signal mixer to avoid local oscillator noise.

§5.11

MODIFIED HIRSHON–FRAENKEL SYSTEM

Hirshon and Fraenkel's system can be modified to allow some of the advantages of the 'balanced mixer' type of spectrometer described in §3.11 to be obtained. The principle of this modification is shown in Fig. 5.9, and consists of a pair of balanced detectors

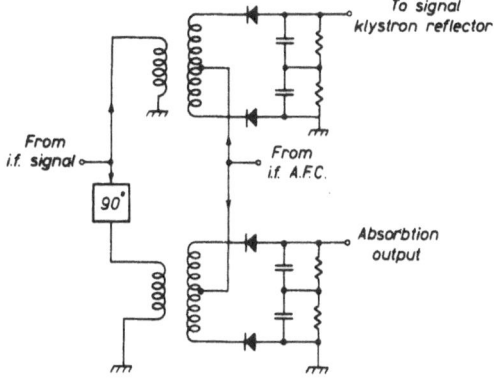

Fig. 5.9. Circuit illustrating the modification to the Hirshon–Fraenkel spectrometer.

which replace the second detector and the phase-sensitive detector of Fig. 5.8. The upper balanced detector simply acts as a phase-sensitive detector and fulfils exactly the function of the phase-sensitive detector of Fig. 5.8. The lower balanced detector, on the other hand, behaves very much like the balanced mixer of §3.11. In that system the microwave bucking signal was fed in at one part of the balanced mixer and the signal from the cavity was fed in at the other. In the present arrangement the two signals are used in the same way, but are first converted to the intermediate frequency. The 90° phase shifting network is needed because the phase of the A.F.C. system responds to reactive reflections from the cavity, whilst, for the absorptive component of the magnetic resonance to be observed, the detecting system must respond to the 'in-phase' reflections. With this method the sensitivity of the spectrometer to frequency drift is reduced for the reasons given in §3.11.

Where dispersion is to be observed, the output from the upper phase-sensitive detector of Fig. 5.9 can be disconnected from the reflector of the signal klystron and used in place of the output from the lower detector.

§5.12

THE SUPERHETERODYNE SYSTEM OF FAULKNER

There seems to be little doubt that the best method of superheterodyne detection that has been devised for a spectrometer so far is the system used by Faulkner.[3] The essentials of Faulkner's spectrometer are shown in Fig. 5.10 and consist of a conventional reflection-cavity spectrometer with a phase-reversing modulator introduced after the output port of the circulator C_1. The modulator consists of a second circulator C_2 and a termination that can be switched electronically ideally from a short

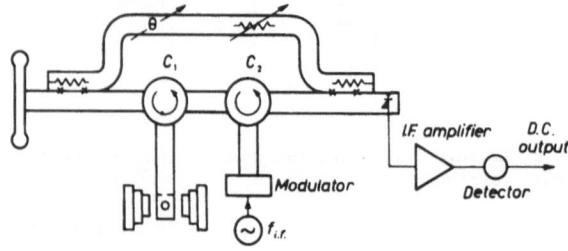

Fig. 5.10. Diagram showing the essential parts of Faulkner's superheterodyne spectrometer.

circuit to an open circuit. The output from C_1 is passed via C_2 to the switched termination, is reflected back to C_2 and out to the detector. In both states of the switch, the modulus of the reflection coefficient of the termination is unity, but the phase of the reflection varies by 180° from one state to the other. When the modulator is driven by an oscillator, its function is, therefore, to reverse periodically the phase of the signal passed from the spectrometer cavity to the detector. The effect of the periodic phase reversal is then that the cavity signal periodically adds to and subtracts from the bucking signal, instead of simply adding to the bucking signal, as in a conventional system. Thus, instead of the cavity signal producing a change in the d.c. output from the detector, an a.c. component is developed with the same frequency as the modulating oscillator and a peak-to-peak amplitude of twice the d.c. change in the conventional system. The effect of

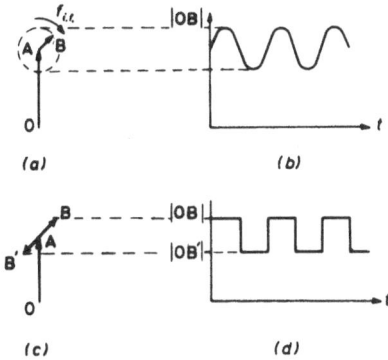

(a)   *(b)*

(c)   *(d)*

Fig. 5.11. Comparison of Faulkner's superheterodyne with the conventional spectrometer. (a) Vector representation of the combination of local oscillator and signal in a conventional superheterodyne, (b) time variation of amplitude of microwave signal reaching mixer corresponding to (a), (c) vector representation of combination of bucking signal OA and switched cavity signal AB, AB' in Faulkner's system, and (d) time variation corresponding to (c).

introducing the modulator is therefore very similar to that obtained by using a local oscillator in the normal type of superheterodyne spectrometer. In both cases the signal from the cavity results in an a.c. signal, instead of a d.c. signal, being developed at the output of the microwave detector. It may be seen from Fig. 5.11 that the sensitivity of the two arrangements are comparable.

In Fig.5.11(a) gives a phasor representation of the operation of a normal superheterodyne spectrometer. Here OA represents the local oscillator signal and AB the signal from the cavity.

Because of the frequency difference between the two, AB rotates relative to OA at the intermediate frequency, and so causes the amplitude of the resultant, which is fed to the microwave detector, to oscillate in the manner shown in Fig. 5.11(*b*). The corresponding action of the modulator spectrometer is slightly different, and the appropriate phasor diagram is that of Fig. 5.11(c). Here, the signal from the cavity, which will not necessarily be in phase with the bucking signal, is periodically reversed in phase and so switches between the values of AB and AB'. The result is to produce a square wave at the output of the detector, because the amplitude of the resultant changes from |OB| to |OB'|. If, then, the phase of the bucking signal is made the same as that from the cavity, the square-wave output from the modulator spectrometer becomes equal in amplitude to the sine-wave output from the normal type of superheterodyne. In both cases the output is passed through the i.f. amplifier, and only the fundamental of the square wave is accepted. However, the amplitude of the fundamental component of a square wave is slightly in excess of that of the square wave itself, so the efficiency of the modulator system is actually a little better than that of the normal superheterodyne. In both cases there is no reason why the intermediate frequency should not simply be that which gives the best overall noise factor (~ 30 Mc/s), and so the modulator spectrometer has a basic sensitivity comparable with that of a normal superheterodyne spectrometer. There is, however, the simplification that only one klystron is needed and the problem of maintaining the correct frequency separation between the two klystrons of the normal superheterodyne does not arise.

The modulator spectrometer can be easily used to distinguish between the absorptive and dispersive components of the magnetic resonance. It is shown that the effect of the modulator is simply to produce an a.c. signal at the output of the microwave detector; this signal has a peak-to-peak amplitude equal to twice the change in the d.c. level produced by the cavity signal from the detector of a conventional spectrometer. The a.c. signal is then amplified and converted to d.c. by the second detector. Thus, the d.c. output from the second detector of the modulator spectrometer is simply an amplified version of the change in d.c. level that the cavity signal produces at the microwave detector of a conventional spectrometer. Thus the mechanism by which the modulator spectrometer distinguishes between absorption and dispersion is exactly the same as that for the conventional spectrometer.

The principle advantage of the modulator spectrometer is, however, that not only is the mechanism for distinguishing between absorption and dispersion the same as for a conventional spectrometer, but so is the process by which f.m. noise on the klystron is converted into a.m. noise. It is shown in §5.2 that the major disadvantage of the normal superheterodyne system is that the efficiency of conversion of f.m. noise to a.m. noise is, for a given degree of mis-tuning of the cavity, increased by the extent to which it is necessary to reduce the level of the effective bucking signal $(\sim \rho_{DO})$. This reduction needs to be rather more than the ratio a_i of the noise factor of the conventional spectrometer to that of the superheterodyne, which for low field modulation frequencies can be considerable. For the modulator spectrometer this problem is avoided because the bucking signal is kept at its full strength. It must be remembered that although the conversion of f.m. noise to a.m. noise is no more severe in a modulator spectrometer than in a conventional spectrometer, it is also no less. Thus, whilst conversion of a conventional spectrometer to the modulator type of superheterodyne detection reduces the basic noise factor of the detecting system, the klystron f.m. noise being unchanged can prevent the overall noise level from being greatly improved. It is therefore clear that to obtain a comparable reduction in the conversion of klystron f.m. noise, the frequency drift of the klystron must be reduced by the factor a_i. This figure compares favourably, however, with the corresponding figure of over a_i^2, which applies for the normal superheterodyne.

In order to obtain an adequate efficiency of detection for the second detector, a small amount of a.c. drive is required at the intermediate frequency. This can be provided either by slightly mis-tuning the cavity or by deriving a small signal from the modulating oscillator.

A worthwhile improvement to the arrangement of Fig. 5.10 is to use the balanced mixer of Fig. 5.3 in place of the single microwave detector of Fig. 5.10. The further reduction in sensitivity in klystron a.m. and f.m. noise outlined in §3.11 is then obtained.

The switched termination used for the modulator is preferably a P.I.N. diode, which when forward biased presents a near short circuit to the microwave radiation and when reverse biased a near open circuit. If a conventional diode is used in this position, a near match is obtained when the diode is biased in the forward direction and a near open circuit when the bias is reversed. Thus the modulator becomes a chopping modulator, that is, a modulator which periodically either allows transmission or prevents it. The depth of modulation so produced is then only one half of that previously obtained, and a loss of 6 dB results.

§5.13

MODIFIED POUND STABILIZER SPECTROMETER

An older version of Faulkner's arrangement is the modified Pound stabilizer[4] of Fig. 5.12. Fundamentally, this a conventional reflection-cavity spectrometer using a magic-tee in place of the circulator. A chopping modulator C_1, in the form of a normal microwave detector, is put in the usual position of the detector, so that the signal normally reaching the detector is alternately reflected and absorbed. Part of the chopped signal reflected from C_1 then reaches the microwave detector C_2. The bucking signal is provided by that part of the output from the klystron that passes directly to C_2, and the phase relationship of the chopped signal from the cavity is adjusted relative to the bucking signal by the microwave shifter in the arm of the magic-tee leading to the chopping modulator. The Pound system differs from Faulkner's system only in details. A chopping modulator is used in place of the reversing modulator because P.I.N. diodes were not available at the time when the Pound system was developed. Now there is no reason why a reversing modulator cannot be used. The other disadvantages are the sensitivity loss of 9 dB due to using a magic-tee instead of a circulator, and the fact that the amplitude of the bucking signal cannot be adjusted.

Faulkner's system and the Pound system can be modified so that the bucking signal, rather than the signal from the cavity, is modulated. In Faulkner's system the reversing modulator is placed in the bucking arm instead of in the cavity arm. In the Pound system the positions of the modulator and detector are

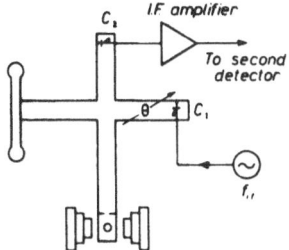

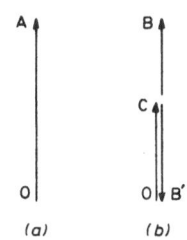

Fig. 5.12. Superheterodyne spectrometer based on the Pound frequency stabilizer.

Fig. 5.13. (*a*) Phasor representing unmodulated output from the klystron. (*b*) Phasor representation of 100 per cent square-wave modulated klystron showing carrier term OC and sideband components CB $\leftrightarrow$ CB$'$ separately.

simply reversed. For the Pound system this modification is in some ways to be preferred because the signal from the cavity is only split once, not twice, before reaching the detector. The amplitude of the bucking signal can then be controlled by inserting a variable attenuator in the arm containing the modulator. A severe disadvantage of modulating the bucking signal, however, is that the amplitude of the reflection coefficient for the two states of the switch must be identical. Any difference will result in the bucking signal being amplitude-modulated at the modulating frequency, even in the absence of a signal from the cavity. When a chopping modulator is used, this effect is gross and the arrangement is not suitable. Moreover, even when the reversing modulator is used, thermal effects tend to cause the amplitude of the two reflection coefficients to drift relative to one another, and there changes are converted directly to changes in the i.f. output. Thus, on the whole, Faulkner's system, as presented in the previous section, is preferable to the Pound system.

§5.14

MODULATED KLYSTRON SUPERHETERODYNE

A very simple way of converting a conventional reflection-cavity spectrometer to superheterodyne operation is to remove the bucking and to amplitude-modulate the signal klystron by switching the grid of the klystron at the required intermediate frequency. The only other requirement then is an i.f. amplifier and detector placed between the output of the microwave detector and the input of the normal amplifying system. The effect of switching the klystron in this way is to generate a series of sidebands separated on either side of the original klystron frequency by multiples of the switching frequency. If the unmodulated output from the klystron is represented by $\hat{e}\sin\omega t$, the effect of the switching is to multiply the unmodulated output by the function $\frac{1}{2} + h_s(t)$, where $h_s(t)$ represents a square wave with a peak-to-peak amplitude of unity and a frequency equal to the required intermediate frequency $f_{i.f.}$. Writing $h_s(t)$ in the form of a Fourier series, we obtain the expression for the modulated output v

$$v = \hat{e}\sin\omega t \left[\frac{1}{2} + \frac{2}{\pi}(\sin\omega_{i.f.}t + \frac{1}{3}\sin 3\omega_{i.f.}t + \frac{1}{5}\ldots \right]$$

$$= \frac{\hat{e}}{2}\sin\omega t + \frac{\hat{e}}{\pi}[\cos(\omega - \omega_{i.f.})t - \cos(\omega + \omega_{i.f.})t] + \text{etc.}$$

$$(5.9)$$

Thus the amplitude of the component at the frequency of the klystron is halved by the process of modulation. If therefore OA in Fig. 5.13(a) represents the unmodulated output from the klystron and (b) represents the modulated output, then OC($= OA/2$) in (b) represents the carrier component of the modulated signal, and the phasor which switches from CB to CB′ in order to provide a resultant which varies between zero and OA, must constitute the sideband components. When the modulated waveform is applied to the spectrometer cavity and circulator, all the sidebands fall at frequencies well outside the resonance curve of the cavity. Therefore the signal represented by the vectors B and B′ is reflected from the cavity and fed to the microwave detector without alteration. The component at the klystron frequency is, however, absorbed by the cavity, since the klystron frequency should coincide with the resonant frequency of the cavity. Furthermore, for other frequencies this system is subject to the complex reflection coefficient of the cavity shown in Fig. 5.2 in the same way as for conventional operation of the spectrometer. Thus the effect of modulating the klystron is very much the same as that obtained by Faulkner's method of §5.12, except that, with the present system, it is the bucking signal that is periodically phase-reversed rather than the signal from the cavity. The main disadvantage of the present system is that the level of the bucking signal which is switched from CB to CB' cannot be altered without influencing the signal from the cavity.

It will be clear that modulating the klystron is not a suitable method for converting a simple transmission-cavity spectrometer to superheterodyne operation, because the sidebands cannot pass through the transmission cavity.

References

1. Friis, H.T., *Proc. I.R.E.*, 1944, **32**, 419.
2. Hirshon, J.M., and Fraenkel, G.K., *Rev. sci. Instrum.*, 1955, **26**, 34.
3. Faulkner, E.A., *Proc. I.E.E.*, 1966, **113**, 1159.
4. Pound, R.V., *Rev. sci. Instrum.*, 1946, **17**, 490.

Chapter 6

Automatic Frequency Control Systems

One of the problems to overcome when designing an E.S.R. spectrometer is to prevent variations in the frequency of the klystron and in the resonant frequency of the spectrometer cavity from occurring. These variations degrade the performance of the spectrometer in one way or another, and various automatic frequency control systems have been devised to prevent this from happening.

In this chapter we shall first list the causes of the variations, explain their effects, and hence determine the general types of stabilizing systems that are required. Then a simple stabilizing system will be described and used as an example to illustrate the general principles of servo theory. After that the simple stabilizer is assessed from the point of view of the stability it can afford, and finally some more advanced types of stabilizer will be described.

§6.1

FREQUENCY DRIFT AND FLUCTUATIONS

The variations in frequency can be divided into two types: drift and fluctuations. The long-term drift in the frequency of the klystron, or that of the cavity, is usually thermal. Either the cavity of the klystron or the spectrometer cavity simply expands, or else the sample heats up owing to the microwave power dissipated in its dielectric losses. Fluctuations in the klystron frequency are partly due to microphonics and partly to the normal f.m. noise of the klystron. If the resonant frequency of the cavity fluctuates, this may be due to microphonics, to vibration of the sample (if it is not adequately clamped) and, what can be most serious of all, to the refrigerant boiling if a tail dewar is placed in the cavity.

168

We shall now show that the main effects of frequency drift are distortion of the X-axis of the recorded spectrum, mixing of the dispersive component of the E.S.R. signal with the absorptive component, and conversion of the frequency fluctuations to noise. The main disadvantage of the frequency fluctuations is that they are a potential source of noise.

Drift in the klystron frequency distorts the X-axis of the spectrum because the condition for magnetic resonance given by equation (1.5) relates the klystron frequency f to the field H_0 of the magnet; the spectrum is traced by varying H_0 and is distorted if f is not constant.

Fig. 3.9 shows how the dispersion signal can be mixed with the absorption signal, and also how the spectrometer responds to the frequency fluctuations. In Fig. 3.9, V_L represents the signal reflected from the cavity of the spectrometer, V_B the bucking signal, and V_D, which is the sum of V_L and V_B, the signal reaching the detector. The absorptive part of the magnetic resonance signal then causes the tip of V_L to move at right angles to the circumference of the circle that V_L describes when the frequency is varied, whilst the dispersive part of the signal causes the tip of V_L to move along the circumference. When the absorption signal is to be observed, the direction of V_D is made perpendicular to the circumference. Then the variation in V_L due to the absorption signal has a first-order effect on the length of V_D, whilst the variation in V_L due to a small dispersion signal produces no significant effect. Alternatively, if the dispersion signal is required, $\dot{V}_D$ is made tangential to the circle. Then the dispersive part of the signal causes $|V_D|$ to change, whilst the absorptive part has no effect.

In either configuration a drift in the frequency of the klystron or of the cavity causes the tip of V_L, and therefore the tip of V_D, to move along the circumference of the circle. If V_D is initially perpendicular to the circle so that absorption is observed, this movement causes the angle between V_D and the circle to change, and the dispersion signal is no longer fully suppressed. If the dispersion signal is required, V_D will initially be tangential to the circle, and then the change in direction of V_D is not so great. Thus the tendency for the absorption signal to mix with the dispersion signal when the dispersion signal is

required is not so great as the tendency for the dispersion signal
to mix with the absorption when the absorption is required.

If either the klystron frequency or the resonant frequency of the
cavity fluctuates, then the tip of V_D fluctuates along the circumference of the circle. The fluctuations therefore have no first-
order effect on $|V_D|$ when the absorption signal is being observed,
and V_D is perpendicular to the circle; but when the dispersion
signal is being observed and V_D is tangential to the circle, the
fluctuations modulate $|V_D|$ with full force. If the frequency
drifts when the absorption signal is being observed, the fluctuations begin to be observed in the same way as the dispersion
signal. When, however, the dispersion signal is being observed,
the fluctuations have the maximum effect initially, and the frequency drift is of little consequence.

We are now a little closer to finding what types of stabilizer
are required when the two types of signal are being observed.
To avoid distortion of the X-axis of the spectrum, the klystron of
both absorption and dispersion spectrometers should be stabilized
relative to some reference frequency that is absolutely constant.
In an absorption spectrometer the relative drift between the frequency of the klystron and that of the spectrometer cavity is
undesirable, because it allows the dispersion signal and the
frequency fluctuations to be observed. Thus there is a need for a
stabilizer that will keep at least the mean frequencies of the
klystron and spectrometer cavity equal. It is less important to
remove the fluctuations about the mean, because these are not
observed if the drift is properly corrected. For the dispersion
spectrometer, a stabilizer is required to remove relative fluctuations between the klystron frequency and the resonant frequency
of the cavity. Here it is less important to correct for relative
drift as the absorption signal does not easily mix with the dispersion signal, and the frequency fluctuations are at their maximum
initially. Indeed the only components of the frequency fluctuations that the stabilizer really needs to correct for are those in
the region of the frequency f_m at which the field of the spectrometer magnet is modulated for double-modulation recording. When
the fluctuations are converted into noise, it is these components
that fall within the band-pass of the spectrometer amplifier
tuned to f_m.

The requirements that we have deduced for the absorption spectrometer stabilizer can all, in principle, be satisfied. It is an easy matter to stabilize the klystron frequency relative to some constant reference frequency, and then the resonant frequency of the cavity can, in principle, be stabilized either to the klystron frequency or to the reference frequency. It is, however, rather difficult to control the cavity frequency because no electrical means of tuning the cavity is immediately available, and such a method of control is not easy — although not impossible — to devise. The klystron frequency, on the other hand, is, via its reflector voltage, readily tunable. Thus, in practice, it is common to have to decide between the two compromise arrangements: (a) stabilizing the klystron frequency to the resonant frequency of the cavity and accepting the X-axis distortion of the spectrum, or (b) stabilizing the klystron frequency to an external reference and accepting the mixing of the dispersion and absorption signals and the conversion of the frequency fluctuations to noise. Usually it is essential to avoid the conversion of noise and the mixing of the signals, so the klystron has generally to be stabilized relative to the cavity.

With the dispersion spectrometer it is not possible, even in principle, to satisfy all of the above requirements. The main difficulty is that the dispersion signal is indistinguishable from a variation in the resonant frequency of the spectrometer cavity; both effects cause the tip of V_D to move along the circumference of the circle, whilst the movement due to the absorption signal is at right angles to it. Thus a stabilizer that is arranged to remove relative variations between the klystron frequency and the resonant frequency of the cavity also removes the dispersion signal. It is therefore only possible for a stabilizer to remove the absolute variations in the klystron frequency, and this must be done by comparing the klystron frequency with some absolute reference frequency; it is no longer possible to stabilize relative to the resonant frequency of the cavity. As it happens, stabilization relative to an absolute reference is necessary to avoid the frequency drift that results in a distortion of the X-axis of the spectrum. Thus to remove the fluctuations at the modulating frequency f_m it is simply necessary to ensure that the frequency response of this stabilizer is wide enough to operate at f_m. Since, in the dispersion spectrometer, the klystron cannot be stabilized relative to the spectrometer cavity, it is just as well that relative drift between the two is not particularly important when dispersion is being observed.

When the klystron is stabilized relative to an external reference, it is generally important that the reference frequency should not drift; but it is also important, particularly when dispersion is being observed, that the reference frequency should not fluctuate, as otherwise the fluctuations will be transferred to the klystron frequency. The question of fluctuations (or, more correctly, of effective fluctuations) in the reference frequency of a stabilizer is an important factor in the assessment of the stabilizer, and is considered in some detail in the following sections of the chapter.

As it is not possible to correct for fluctuations in the resonant frequency of the cavity when the dispersion signal is being observed, and since the spectrometer then responds to fluctuations to the fullest extent, it is very important, in this instance, that frequency fluctuations should be prevented from occurring. A useful evasive measure that can be taken if the fluctuations cannot be prevented, is to make f_m as high as possible. Then, because the spectral intensity of the frequency fluctuations usually decreases with frequency, and because the spectrometer amplifier is tuned to f_m, the amplifier accepts noise over a region where the spectral intensity of the fluctuations is reduced.

§6.2

SIMPLE D.C. FREQUENCY STABILIZER

We shall now describe one of the simplest methods of stabilizing the klystron frequency. The arrangement is shown in Fig. 6.1 and is not dissimilar to an E.S.R. spectrometer. The microwave section of the stabilizer forms what is known as a simple microwave discriminator. The function of the discriminator is to preduce a voltage at the output of the balanced detector $D_1 D_2$ that is proportional to the difference between the frequency of the klystron K and the resonant frequency of the reference cavity

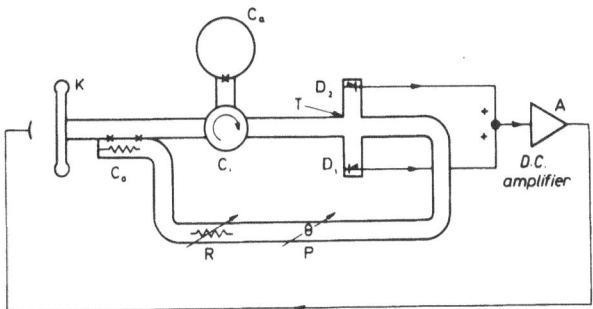

Fig. 6.1. Simple microwave discriminator for stabilizing the klystron frequency to the resonant frequency of a microwave cavity.

C_a. The error voltage is then amplified by the d.c. amplifier A and fed to the klystron reflector, with the appropriate polarity, to correct for the original error in frequency.

The discriminator operates in the following manner. The output from the klystron is divided by the directional coupler C_0, most of the power going via the circulator C_i to the cavity C_a. The signal reflected from the cavity then goes again via the circulator to the magic-tee T. There it is divided between the two halves of the balanced detector D_1 and D_2. The remainder of the output from the klystron passes via the side arm of the coupler C_0, the variable attenuator R, and the phase-shifter P, to the remaining port of the magic-tee. There it is also divided to combine with the signal from the cavity in the two halves of the detector. The manner in which the signals combine is shown in Figs. 6.2(a) and (b). Here (a) represents the combination at D_1, and (b) the combination at D_2. We shall consider first the detector D_1. V_{L1} in (a) represents the signal from the cavity at D_1, and V_{B1} the bucking signal. $V_{D1} = V_{L1} + V_{B1}$ and is the combined signal reaching D_1. From §3.2 it will be seen that the tip of V_{L1}, and therefore of V_{D1}, traces the circle shown when the klystron frequency f varies through the resonant frequency f_0 of the cavity. The cavity is matched so that the circle passes through the origin when $f = f_0$. The phase relationship of V_{B1} is adjusted by the phase-shifter P. For the phase relationship shown, $|V_{D1}|$ varies with the klystron frequency as shown in Fig. 6.2(c). The detector output therefore varies linearly with

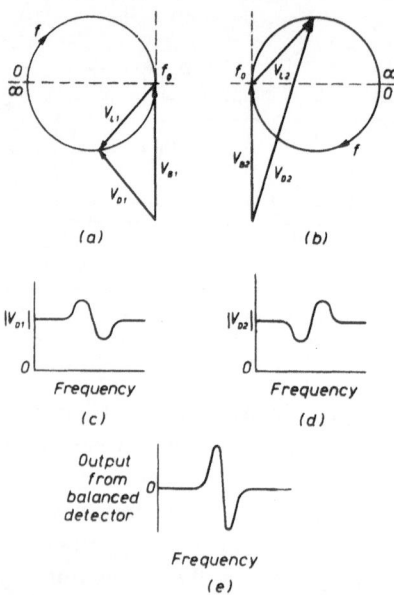

Fig. 6.2. Action of a microwave discriminator.
(a) Vector representation of bucking V_{B1} and cavity V_{L1} signals fed to detector D_1.
(b) Vector representation of bucking V_{B2} and cavity V_{L2} signals fed to detector D_2.
(c) Amplitude of combined input to D_1 as a function of frequency.
(d) Amplitude of compined input to D_2 as a function of frequency.
(e) Output from balanced detector as a function of frequency.

frequency as f passes through f_0, but is not equal to zero when $f = f_0$. The standing d.c. level is removed by using the second half of the detector. For D_2, Fig. 6.2(b) differs from (a) because the magic-tee splits the phase of one or the other of the two inputs. Here we assume that it is the signal from the cavity, not the bucking signal, that is split (the principle would be unaltered if the reverse were the case). Then $|V_{D2}|$ varies as shown in (d). The two detectors are arranged to be of opposite polarity, so the combined output, shown in (e), is proportional to $|V_{D1}| - |V_{D2}|$. The standing d.c. level is thus cancelled and the output discriminator is truly proportional to the frequency error.

§6.3

INCORPORATION OF THE SIMPLE STABILIZER
INTO A SPECTROMETER

The simple frequency stabilizer of the previous section can be used to stabilize the klystron frequency in an E.S.R. spectrometer either to some external reference frequency or to the resonant frequency of the spectrometer cavity. We shall consider stabilization to the reference frequency first. Here the usual arrangement is to use a side-arm coupler to tap-off a fraction of the output from the spectrometer klystron, and to use the tapped-off output to drive a sample microwave discriminator that is built around a separate reference cavity. The cavity is tuned to the desired frequency. The d.c. output from the discriminator is amplified and applied to the klystron reflector in the usual way. The fraction of power used to drive the discriminator is usually about one tenth of the total output from the klystron. The power reaching the cavity of the spectrometer is thereby not significantly altered, whilst at the same time enough power is provided for the discriminator to work. A 10 dB coupler is used to achieve this fraction.

It is possible to use the simple stabilizer to stabilize the klystron frequency to the resonant frequency of the spectrometer cavity if the spectrometer happens to be the balanced mixer spectrometer of §3.11 or its near equivalent. Comparing Figs. 6.1 and 3.20 shows that the microwave circuit for the stabilizer is exactly the same as that for the spectrometer. Thus it would seem possible to split the signal at the output of the balanced detector, feeding the a.c. component to the spectrometer amplifier and the d.c. component to the A.F.C. amplifier. This unfortunately cannot be done because the phase relationship of the bucking signal used with the absorption spectrometer (and only when the absorption signal is being observed is it permissible to stabilize the klystron frequency to the resonant frequency of the spectrometer cavity) differs by 90° from that used with the discriminator. The discriminator is phased to respond to frequency changes and thus will respond to the dispersion signal but not to the absorption signal.

The arrangement needed to overcome this problem is shown in

Fig. 6.3. The system here is basically a spectrometer, with D_1 and D_2 forming the spectrometer detector. The bucking signal however is split at the magic-tee T_2, and half of the power goes to an A.F.C. detector D_3 and D_4. The signal for the A.F.C. detector is derived from the directional coupler C_2. In this way

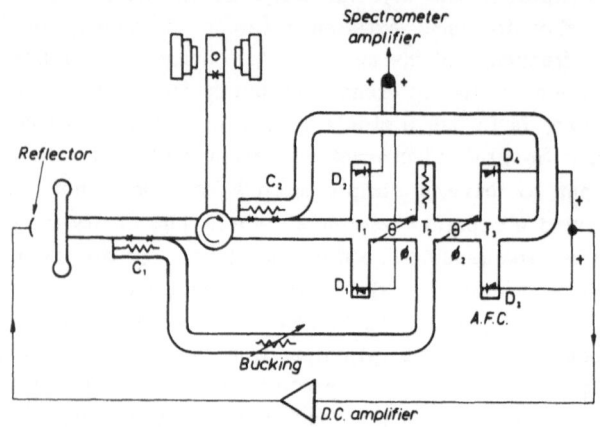

Fig. 6.3. Spectrometer with a simple microwave discriminator incorporated to stabilize the klystron frequency to the resonant frequency of the spectrometer cavity.

the phase relationship of the bucking signals to the A.F.C. and the spectrometer detectors can be controlled independently by the phase shifters ϕ_1 and ϕ_2. Thus ϕ_2 can be adjusted so that the A.F.C. detector responds to frequency variations, whilst ϕ_1 can be adjusted so that the spectrometer detector responds to the absorption signal. The coupling factor of C_2 is made 10 dB, because this does not rob the spectrometer detector of too large a fraction of the E.S.R. signal and still enables the stabilizer to function.

§6.4

ELEMENTS OF SERVO THEORY

A frequency stabilizer will completely remove variations in the klystron frequency only if the loop gain of the stabilizer is infinite. Otherwise a finite frequency error has to be used to develop the correcting voltage applied to the klystron reflector.

We now introduce the elements of standard servo theory[1] that relate to this consideration. Here it will be shown how the maximum permissible loop gain is determined by the frequency response of the stabilizer.

In general, the stabilizer reduces an error δf_{KU} in the frequency of the unstabilized klystron to the value δf_{KS}, where δf_{KS} is related to δf_{KU} by the following equation.

$$\delta f_{KS} = \delta f_{KU}[1 + kG(\omega)]^{-1} \qquad (6.1)$$

Here $kG(\omega)$ is the loop gain of the stabilizer and is a function of frequency. The simple stabilizer is used as an example from which (6.1) can be deduced. First, the arrangement of Fig. 6.1 is re-drawn in the form of Fig. 6.4. In Fig. 6.4, k_{DIS} represents

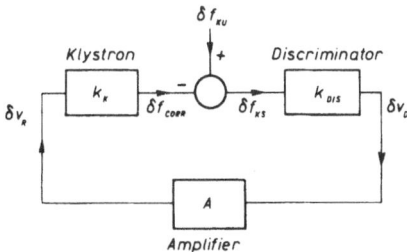

Fig. 6.4. Schematic representation of the system of Fig. 6.1 for servo theory analysis.

the response of the discriminator, and relates the output voltage δv_D from the detector of the discriminator to the error in the klystron frequency that causes it. With the present arrangement the frequency error is equal to δf_{KS}, so

$$\delta v_D = k_{DIS}\delta f_{KS} \qquad (6.2)$$

A represents the voltage gain of the d.c. amplifier, i.e.

$$\delta v_R = A\,\delta v_D \qquad (6.3)$$

where δv_R is the change in voltage applied to the klystron reflector. k_K relates the change in the klystron frequency to δv_R (the change is due to δv_R). It is this frequency change, termed δf_{CORR}, that corrects for the original change δf_{KU}. Thus

$$\delta f_{CORR} = k_K\delta v_R \qquad (6.4)$$

and

$$\delta f_{KS} = \delta f_{KU} - \delta f_{CORR} \qquad (6.5)$$

It is therefore appropriate to complete the loop by the differencing junction that represents (6.5). Equations (6.2) to (6.5) can be combined to give

$$\delta f_{KS} = \delta f_{KU}(1 + k_{DIS}A k_K)^{-1} \qquad (6.6)$$

But $k_{DIS}A k_K$ is the loop gain of the stabilizer, so (6.1) follows directly.

It is clear that the loop gain should be as high as possible. In practice, however, one finds that continued advancement of the loop gain always eventually results in a form of oscillation known as 'hunting'. We shall now show that the maximum loop gain that can be used before hunting starts is determined by the frequency response of the stabilizing loop.

Each of the factors k_{DIS}, A and k_K that make up the loop gain are functions of frequency, that is, when the input to any of these factors is sinusoidal, and the frequency of the sine wave is increased, there comes a point where the corresponding output decreases and varies in phase relative to the input. It is usual to use k_A to represent the low-frequency value of the loop gain and $G(\omega)$ to represent the fall-off as (ω) is increased. The low-frequency value of $G(\omega)$ is therefore 1. It happens, in the present system, that the only factor that depends significantly on frequency is A. We therefore write A as $k_A G_A(\omega)$ and determine $G_A(\omega)$ from Fig. 6.5. Fig. 6.5 is a block-diagram representation of a typical arrangement for A (in fact, a four-stage d.c. amplifier) with the elements determining $G_A(\omega)$ shown specifically. In the diagram, R_0 is the output impedance of each stage, R_i the input impedance, and C_i the input capacity. We represent the gain of each stage, say the rth, by the factor $k_{AR}G_{AR}(\omega)$. Then by simple network theory

$$k_{AR}G_{AR}(\omega) = k_{AR}\left(1 + \frac{j\omega}{\omega_R}\right)^{-1} \qquad (6.7)$$

where $\omega_R = (R_i + R_0)/C_i R_i R_0$. The end networks are slightly different and give expressions of the same form but with values of ω_R equal to $\omega_1 = (R_i + R_D)/C_i R_i R_D$ for the first network, and $\omega_5 = (C_K R_0)^{-1}$ for the last. Here R_D represents the output

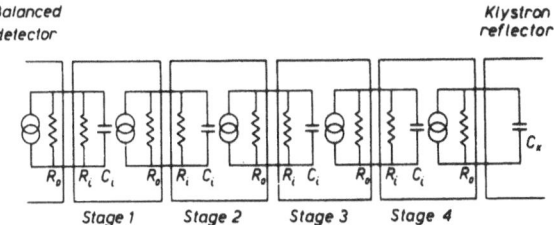

Fig. 6.5. Block diagram representing the amplifier in Fig. 6.1.

impedance of the balanced detector, and C_K represents the input
capacity of the reflector of the klystron. Then

$$k_A G_A(\omega) = k_{DIS} k_A k_K \left[\left(1 + \frac{j\omega}{\omega_1} \right) \cdots \left(1 + \frac{j\omega}{\omega_5} \right) \right]^{-1} \quad (6.8)$$

By standard a.c. theory, the combined phase shift ϕ_A of the
loop is given by

$$\phi_A = \sum_{R=1}^{5} \phi_{AR}$$

where

$$\phi_{AR} = -\tan^{-1}\left(\frac{\omega}{\omega_R}\right) \quad (6.9)$$

The combined attenuation, represented by $|G_A(\omega)|$, is given by

$$|G_A(\omega)| = \left\{ \left[1 + \left(\frac{\omega}{\omega_1}\right)^2 \right] \left[1 + \left(\frac{\omega}{\omega_2}\right)^2 \right] \cdots \right\}^{-\frac{1}{2}} \quad (6.10)$$

We shall now assume that ω_1 to ω_5 are all equal (this, it
will shortly be seen, is the condition most favourable for hunting
to occur) and designate the common cut-off frequency ω_c. Then
the attenuation and phase-shift, from equations (6.10) and (6.9),
are as shown in Figs. 6.6(a) and (b). Further, the combined
attenuation and phase shift for the whole loop is obtained by
multiplying the curves of (a) and (b) by 5, which gives the
curve of (d). Then, from (c) and (d) it is seen that the total
phase-shift reaches and passes through a value at 180° at a frequency a little below ω_c. At this frequency the phase of the
feedback is therefore reversed, and any variation in the klystron
frequency is enhanced rather than stabilized. If the modulus of
the loop gain exceeds unity at this frequency, the loop oscillates;

this is 'hunting'. If however the loop gain is made less than unity at this frequency, the gain is never enough at any frequency to be significant. At lower frequencies $|G(\omega)|$ increases slightly, but is not enough to be of any use.

The way to overcome this problem is to reduce considerably the cut-off frequency of *one* of the networks. In the present example this would probably be arranged by shunting C_K by an additional capacitor, and adding a resistance between the output of the amplifier and the klystron reflector. The value of ω_s is thus reduced to a value of ω_s' and gives the overall gain and phase response shown in Figs. 6.5(*e*) and (*f*). Then, by the time the frequency at which the phase lag of 180° is reached, the value of $|G(\omega)|$ has fallen by almost the ratio ω_s'/ω_c. Thus, the setting of gain which gives a value of unity for the loop gain at the frequency of phase reversal, gives a low-frequency loop gain of nearly ω_c/ω_s'. From this argument it is clear that the maximum degree of stabilization that can be obtained is given roughly by the ratio of the lowest cut-off frequency in the loop to the next lowest. For the present example the added resistance might be 1 MΩ and the added capacity 10 μF. Because these values would be much greater than those of the components they supplement, the value of $\omega_s' = (CR)^{-1}$ is then $\simeq 0\cdot 1$ rad/sec. Typically the d.c. amplifier might have a cut-off frequency of $100k$ rad/sec, so, in principle, a zero loop gain k of approaching 10^6 can be obtained.

Since it is advantageous to have one cut-off frequency within the response of the servo loop, which is much lower than any of the others, the ideal arrangement includes an integrator in the loop. In effect, an integrator can be considered as a low-pass network with an infinitely low cut-off frequency. If the input to an integrator is of the form $a\cos\omega t$, then the output is $a/\omega \times \cos(\omega t - 90°)$. Thus, the amplitude of the output is inversely proportional to the frequency of the input, and the phase-lag is always 90°. Inspection of equations (6.7) and (6.9) or of Figs. 6.6(*a*) and (*b*) will show that this is the response of a single low-pass network at frequencies well above its cut-off frequency. When such a characteristic is added to the response of Figs. 6.6(*c*) and (*d*), the effect is to lower all the phase-shift values by a further 90° and to modify the gain response so that

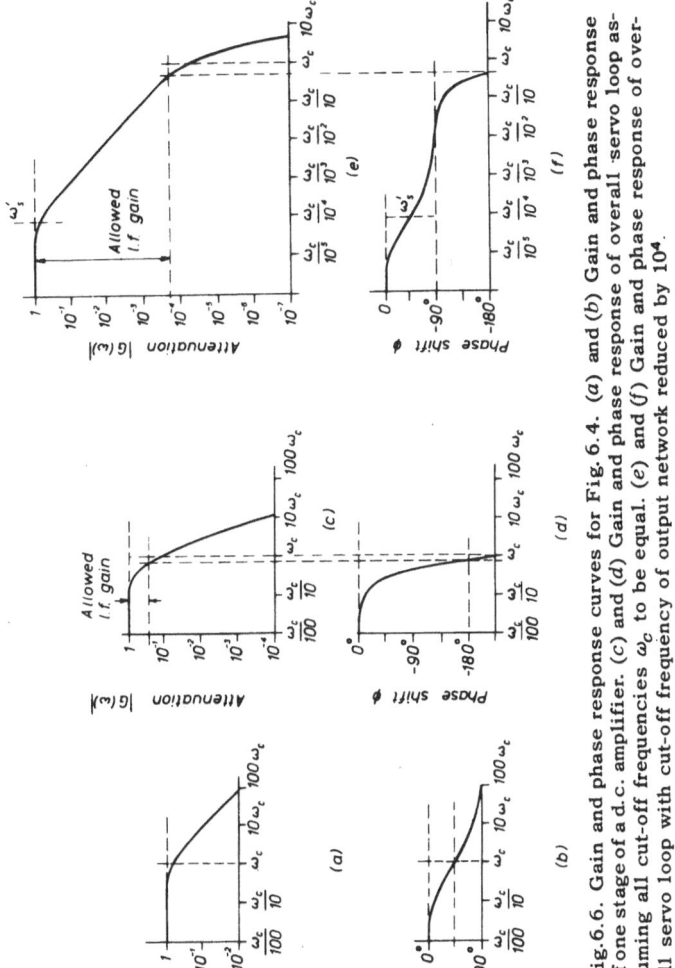

Fig.6.6. Gain and phase response curves for Fig. 6.4. (a) and (b) Gain and phase response of one stage of a d.c. amplifier. (c) and (d) Gain and phase response of overall 'servo loop assuming all cut-off frequencies ω_c to be equal. (e) and (f) Gain and phase response of overall servo loop with cut-off frequency of output network reduced by 10^4.

$|G(\omega)| \sim \omega^{-1}$ for frequencies well below the cut-off frequency and $|G(\omega)| \sim \omega^{-6}$ for frequencies well above the cut-off frequency; previously, $|G(\omega)|$ was constant for $\omega \ll \omega_c$ and $|G(\omega)| \sim \omega^{-5}$ for $\omega \gg \omega_c$. The overall loop gain then needs to be reduced slightly, because the gain must be limited to unity at the frequency where the phase shift was previously 90° rather than 180°. On the other hand, instead of the loop gain settling to a value of only 1 or 2 for all frequencies below the frequency of phase reversal, the gain rises $\sim \omega^{-1}$ and continues to do so indefinitely. Thus, rather than the gain stabilizing at a value of ω_c/ω_5' when one of the cut-off frequencies is lowered, it rises to a zero frequency value of infinity. When an integrator is included the steady-state frequency error is therefore completely removed.

A servo which incorporates an integrator into the control loop is known as a class 1 servo, whilst the previous arrangement is a class 0 servo. In fact, the class of the servo is simply equal to the number of integrators in the loop.

One type of integrator, which is sometimes used in practice, is the electronic integrator. On close inspection, however, an electronic integrator is nothing more than a low-pass network with a much reduced cut-off frequency. Hence a system incorporating an electronic integrator is really only a class 0 system. Another type of integrator is the servo motor, where, ideally, the angular displacement of the motor is proportional to the integral of the input signal. A motor can be utilized in the present system by feeding the output from the d.c. amplifier to the motor, and using the motor to control the 'reflector voltage' knob on the klystron power supply. The system then ideally has an infinite gain at zero frequency, since, for a frictionless motor, if one waits long enough, the motor continues to turn until the error is completely removed. The effects of friction tend, of course, to restore a little of the error.

§6.5

ASSESSMENT OF SIMPLE STABILIZER

It is now important to ask to what extent do the frequency drift and fluctuations obtained with the simple d.c. frequency stabilizer degrade the performance of the E.S.R. spectrometer with which the stabilizer is used. We shall first find the extent of the drift and fluctuations, and then relate the drift to the resulting distortion of the X-axis of the recorded spectrum and to the extent to which the dispersion signal is mixed with the absorption signal in an absorption spectrometer. For the absorption spectrometer we shall also determine the extent to which the frequency fluctuations are converted to noise. The converted frequency modulation noise is then compared with the existing noise from the spectrometer. For the dispersion spectrometer the noise resulting from the fluctuations is also calculated and compared with the existing spectrometer noise.

Drift

We shall start by determining the frequency drift. This generally originates from d.c. drift in the output from the detector of the stabilizer and also from the d.c. amplifier. The d.c. drift is converted to a frequency error because the drift signal is amplified and fed to the klystron reflector. If the loop gain is high, this results in the klystron frequency being changed enough to produce an output from the discriminator that all but cancels the drift voltage. It is therefore clear that, for a given d.c. drift, the sensitivity of the discriminator should be high. The particular microwave arrangement of Fig. 6.1 is chosen for this reason. The required discriminator circuit must give the greatest change in the signal reflected from, absorbed by, or transmitted through, a cavity for a given change in frequency. The dispersion component of an E.S.R. signal has the same effect as a change of frequency, so the spectrometer arrangement that is most sensitive for detecting dispersion signals should also form the best discriminator. It is shown in §2.5 that the most sensitive circuit is the reflection-cavity-circulator arrangement of Fig. 2.16, so this is the circuit that is used for the discriminator, using however a balanced detector.

The single-ended detector arrangement of Fig. 2.16 can be used for the discriminator as long as the bucking signal has the appropriate phase relationship. It is then necessary to balance out the standing d.c. potential from the output of the detector, which can be done by a suitable d.c. source. The main disadvantage of this arrangement is that drift in the amplitude of the output from the klystron adds to the d.c. drift; with the balanced detector this drift is cancelled.

The frequency drift is related to the d.c. drift by determining what change in the klystron frequency is required to cancel the d.c. drift voltage. It turns out to be most convenient to relate the frequency drift to the fractional change $\delta|V_B|/|V_B|$ in the amplitude of the bucking voltage V_B to the detector that would be needed to cancel the d.c. drift. Then δf, the resulting frequency error, is given by the expression

$$\delta f = (\Delta f)_U \left(\frac{P_B}{P_C}\right)^{\frac{1}{2}} \frac{\delta|V_B|}{|V_B|} \qquad (6.11)$$

where $(\Delta f)_U$ is the unloaded bandwidth of the cavity, P_B the bucking power, and P_C the power dissipated in the cavity. This is a useful way of expressing the result, because $\delta|V_B|/|V_B|$ can easily be measured or, indeed, guessed.

The value of $\partial V_L/\partial X$, the rate of change of the load voltage with cavity reactance, is given for the reflection-cavity spectrometer by equation (3.13) as $-jE_g/4R$, where R is the effective resistance of the cavity and E_g the generator voltage associated with P_C. From §3.4 $X = 2\delta f R/(\Delta f)_U$ so

$$\delta V_L = \frac{-jE_g}{2(\Delta f)_U} \delta f \qquad (6.12)$$

With the phase relationship for discriminator operation, it is clear from §6.2 that $d|V_D|/df = |dV_L/df|$, so δf can be related to $\delta|V_D|$, the change in the amplitude of the input voltage to the detector that cancels the drift voltage.

$$\delta f = \frac{2(\Delta f)_U}{|E_g|} \delta|V_D| \qquad (6.13)$$

Now $|E_g|^2/8R_0 = P_C$ and $|V_B|^2/2R_0 = P_B$, where R_0 is the waveguide impedance, so

$$\delta f = (\Delta f)_U \left(\frac{P_B}{P_C}\right)^{\frac{1}{2}} \frac{\delta|V_D|}{|V_B|} \qquad (6.14)$$

But $\delta|V_D|$ could be due to a change $\delta|V_B|$, so equation (6.11) follows.

The factor $\delta|V_D|$ in (6.13) depends on the bucking power, and can be written as

$$\delta|V_D| = \left(\frac{dv_D}{d|V_D|}\right)^{-1} \delta v_D \qquad (6.15)$$

where δv_D is the change in output voltage from the detector due to $\delta|V_D|$. $dv_D/d|V_D|$ is the slope of the microwave input, d.c. output characteristic of the detector, whilst δv_D is the equivalent drift voltage. Thus $dv_D/d|V_D|$ and δv_D figure in equation (6.15) in exactly the same way as the (conversion gain)$^{\frac{1}{2}}$ and the r.m.s. noise voltage figure in equation (2.20) for the overall noise factor of the detector. Also, δv_D will probably vary with the bucking power in about the same way as the noise does. Thus it is likely that the bucking power that gives the best noise factor will also give the lowest value of $\delta|V_D|$. From Fig.2.24(a) this value is 0·5 mW.

For the optimum bucking power, a probable value of $\delta|V_B|/|V_B|$ in equation (6.11) would be about 1 per cent. Then, if, say, $(\Delta f)_U = 3$ Mc/s, $P_B = 0\cdot5$ mW and $P_C = 50$ mW, δf would be 3 kc/s.

The X-axis distortion of the spectrum that results from a drift of this amount is very small. If the operating frequency of the spectrometer is 10 kMc/s, the resulting fractional drift is 3 parts in 10^7. The fractional magnetic resonance linewidth seldom approaches this value.

Admixture of dispersion

We shall now determine the extent to which a given frequency drift allows the dispersion signal to be mixed with the output from an absorption spectrometer. To this end we define a dispersion suppression factor S, which is the ratio of the observed dispersion signal to the dispersion signal observed by the corresponding dispersion spectrometer. As the dispersion signal is potentially of the same magnitude as the absorption signal, S is also the ratio of the observed dispersion signal to the observed absorption signal.

When determining S as a function of δf, its value depends on whether a balanced or single-ended detector is used with the spectrometer. This consideration was introduced somewhat qualitatively in §3.11, where it was shown that the balanced mixer lowered the conversion of frequency fluctuations to noise, and also reduced the tendency for the dispersion and absorption signals to mix. In determining the value of S for both a single-ended and a balanced mixer, it is now possible to justify this conclusion quantitatively, and it transpires to qualify the conclusion somewhat. We shall now show that for the single-ended detector

$$S = \left[\left(\frac{P_C}{P_B} \right)^{\frac{1}{2}} \pm 2 \right] \frac{\delta f}{(\Delta f)_U} \qquad (6.16)$$

and for the balanced detector

$$S = 2 \frac{\delta f}{(\Delta f)_U} \qquad (6.17)$$

where the sign of (6.16) depends on the phase relationship of the bucking signal.

Fig. 6.7 can be considered to represent the combination of signals at the input of the single-ended detector, where V_L is assumed to result from the frequency drift δf. Since the component of the dispersion signal observed is the component parallel to V_D, it is clear that $S = \sin \phi$. But if $|V_L| \ll |V_B|$, $\sin \phi \simeq \phi$. Now $\phi = \phi' + \phi''$, and $\phi' \simeq |V_L| / |V_B|$ and $\phi'' \simeq |V_L| / R_c$, where R_c is the radius of the circle. Thus

$$S = \frac{|V_L|}{|V_B|} + \frac{|V_L|}{R_c} \qquad (6.18)$$

If the phase of V_B is reversed, it can similarly be shown that

$$S = \frac{|V_L|}{|V_B|} - \frac{|V_L|}{R_c} \qquad (6.19)$$

Now $R_c = |E_g|/4$, $|E_g|^2/8R_0 = P_C$ and $|V_B|^2/2R_0 = P_B$, so, with equation (6.12), equation (6.16) follows.

Equation (6.19) is an interesting result because it suggests that when $P_C = 4P_B$, S is 0 whatever the frequency drift. This happens because, when $|V_B| = R_c$, V_D joins the centre of the circle to its circumference, and so remains perpendicular to the circumference whatever the value of δf. In cases of severe frequency drift or fluctuations this is a convenient way to operate the spectrometer.

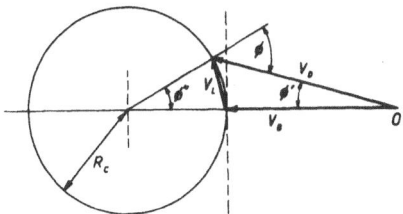

Fig. 6.7. Phasor diagram for determining the dispersion suppression factor S.

If a balanced detector is used,[1] the two versions of equation (6.16) apply to the two halves of the detector, since the phase of either V_L or V_B is reversed for the second half of the detector. The two detectors are of opposite polarity so the difference between the two values is taken. Simply taking the difference gives a result equal to twice that given by equation (6.17). The full justification for the adjustment is slightly complicated: the signal V_L is divided between the two halves of the detector, the two signal outputs from the detectors are correlated, whilst the two noise powers are not, etc. (see §5.7). It will probably be sufficient to note that the adjustment is needed if the maximum value of S is to equal one.

To determine the value of S obtained with the simple stabilizer, the value of δf given by equation (6.11) must be inserted into (6.16) and (6.17). This gives

$$S = \left[\left(\frac{P_{CE}}{P_{BE}}\right)^{\frac{1}{2}} \pm 2\right]\frac{(\Delta f)_{UA}}{(\Delta f)_{UE}}\left(\frac{P_{BA}}{P_{CA}}\right)^{\frac{1}{2}}\frac{\delta|V_B|}{|V_B|} \qquad (6.20)$$

for the single-ended detector and

$$S = 2\frac{(\Delta f)_{UA}}{(\Delta f)_{UE}}\left(\frac{P_{BA}}{P_{CA}}\right)^{\frac{1}{2}}\frac{\delta|V_B|}{|V_B|} \qquad (6.21)$$

for the balanced detector. In these equations the suffix E is used to refer to the E.S.R. spectrometer, and the suffix A to the A.F.C. system (the stabilizer). For an absorption spectrometer it is normal to stabilize the klystron to the resonant frequency of the cavity (using the arrangement of Fig. 6.3) and then the distinction does not need to be made. However, the coupling factor α_2 (α_2 = power loss) of the coupler C_2 needs to be taken into account, and so we shall multiply δf in equation (6.11) by $a_2^{\frac{1}{2}}$. Thus for the composite spectrometer-stabilizer

$$S = \left[1 \pm 2\left(\frac{P_B}{P_C}\right)^{\frac{1}{2}}\right] a_2^{\frac{1}{2}} \frac{\delta|V_B|}{|V_B|} \qquad (6.22)$$

if a single-ended detector is used with the spectrometer, and

$$S = 2\left(\frac{P_B}{P_C}\right)^{\frac{1}{2}} a_2^{\frac{1}{2}} \frac{\delta|V_B|}{|V_B|} \qquad (6.23)$$

if the balanced detector is used. If stabilization is to an external reference cavity, (6.20) and (6.21) remain appropriate. When the external cavity is used, the primary cause of drift is probably the relative drift between the spectrometer cavity and the reference, so there is little point in calculating the drift due to δf. However, with the more usual arrangement, $\delta|V_B|/|V_B|$ might be 1 per cent, $a_2 = 10$, $P_B = 0\cdot5\,\text{mW}$ and $P_C = 1\,\text{mW}$. Then S is in the region of 5 per cent for both balanced and unbalanced detectors.

We shall now determine the extent to which the frequency drift allows the frequency fluctuations to be converted to noise in an absorption spectrometer. This is easily done by determining the relationship between the frequency fluctuations and the resulting noise in a dispersion spectrometer (this relationship will be required later for the dispersion spectrometer). As the factor S is known and as the frequency fluctuations are subject to S in exactly the same way as the dispersion signal, the result for the dispersion spectrometer can simply be multiplied by S.

We shall now show that, for the dispersion spectrometer, the noise power P_n' at the input of the detector of the spectrometer due to the frequency fluctuation is related to the mean square value $\overline{\delta f^2}$ of the frequency fluctuation by the equation

$$P_n' = \frac{P_{CE}}{(\Delta f)_{UE}^2} \overline{\delta f^2} \qquad (6.24)$$

The dispersion spectrometer is equivalent to the stabilizer, so the change $\delta|V_D|$ in detector voltage for a given frequency error δf is given by equation (6.13). The resulting power P_n' to the detector is therefore $\delta|V_D|^2/2R_0$, and $P_C = |E_g|^2/8R_0$, so (6.24) follows.

P_n' must be compared with the normal noise from the

spectrometer P_n. Now $P_n = F_E kTB$, where F_E is the noise factor of the spectrometer, and B the bandwidth over which the noise (and hence the fluctuations) is considered. Thus if the spectrometer is operated in the dispersion mode,

$$\frac{P_n'}{P_n} = \frac{P_{CE}}{(\Delta f)_{UE}^2 \, F_E \, kTB} \, \delta f^2 \qquad (6.25)$$

and when absorption is being observed,

$$\frac{P_n'}{P_n} = S^2 \frac{P_{CE}}{(\Delta f)_{UE}^2 \, F_E \, kTB} \, \delta f^2 \qquad (6.26)$$

The value of δf^2 will depend on whether the fluctuations are stabilized or not. We shall take the pessimistic view here and imagine that they are not. If they are, the value can be obtained from equation (6.28) below. The values of δf^2 for an unstabilized klystron are usually quoted by the manufacturer for a bandwidth of $1 \, \mathrm{kc/s}$ and are, of course, a function of the centre frequency of the band-pass. Typical values are shown in Table 6.1, which is compiled from figures published by Varian Associates for the klystron VA-510B. The values of F_E shown in the table are taken from the measured values of Fig. 2.24(a). Using these values, the value of S required for P_n to equal P_n' is calculated and tabulated. It would thus appear that S is not important since it can never, in fact, exceed unity. However, the values are quoted for a particularly low-noise klystron, and for a value of P_{CE} that is lower than is sometimes used. In addition, the fluctuations in the resonant frequency of the cavity must be allowed for. It is usual to find in practice that the increase in noise becomes significant at about the same value of S as that at which the dispersion signal begins to be troublesome.

For the simple discriminator the expected value of S is about 5 per cent, and this is a little inadequate. Thus for the absorption spectrometer it is usual to adopt one of the more advanced stabilizing systems described in the remainder of this chapter. However, it is possible that the system may be adequate if care is taken to reduce drift in the d.c. amplifier, and if the two halves of the balanced detector are kept in close thermal contact so that differential drift is avoided between the detectors.

Equations (6.16) and (6.17) show that the claims of §3.11

Table 6.1. Variation of permissible dispersion suppression factor S with field modulation frequency f_m.

f_{m}(kc/s)	F_E	$(\overline{\delta f^2})^{\frac{1}{2}}$	S
1	16000	17	1·4
10	1000	3	2
100	63	1	1·6

F_E = noise factor of an E.S.R. detecting system at the modulating frequency f_m

$\overline{\delta f^2}$ = mean square frequency fluctuation of a commerical klystron (Varian VA-510B) over a bandwidth of 1 kc/s and at the frequency f_m

S = hypothetical dispersion suppression factor of the spectrometer needed to make the converted frequency fluctuations of the klystron double the existing noise level.

regarding the balanced mixer spectrometer must be qualified slightly. The dispersion (and noise) suppression factor S is only significantly better for the balanced detector when the cavity power P_C is high ($P_C > 4P_B$); for low values of P_C the values of S are the same for both types of detector. This is not a particularly important point, at any rate from the point of view of noise, because P_{CE} enters into equation (6.26) as well as S. It is therefore at high values of P_{CE} that the improvement is most needed.

We shall now determine the frequency fluctuations obtained with the simple frequency stabilizer, and thence the extent to which they increase the noise level of the associated E.S.R. spectrometer. One obvious cause of frequency fluctuations is fluctuations in the resonant frequency of the reference cavity. If the loop gain of the stabilizer is high, these are transferred to the klystron frequency. If the loop gain is not high (i.e. $kG(\omega) < 1$) the fluctuations are not transferred, nor are the original fluctuations in the klystron frequency removed. There is therefore the general rule that, for frequencies where $kG(\omega) \gg 1$, the fluctuations in the frequency of the unstabilized klystron are exchanged for those in the reference frequency, whilst for higher frequencies, where $kG(\omega) \ll 1$, the original fluctuations in the klystron frequency remain. Matters are complicated however, because noise from the detecting system of the stabilizer also contributes to the frequency fluctuations. This happens in the same way that d.c. drift in the detecting system becomes frequency drift. The klystron frequency is altered enough to produce a voltage at the output of the discriminator that cancels the drift or noise voltage. The above rule can still be used if it is simply imagined that the noise voltage is produced by fluctuations in the reference frequency. Then, when $kG(\omega)$ is high, the fluctuations transferred to the klystron are the real, plus the imagined, fluctuations in the reference frequency.

It is quite simple to determine the imagined fluctuations. Equation (6.24) relates a fluctuation δf in the klystron frequency of a dispersion spectrometer to the resulting noise power P_n' at the detector. The stabilizer is the same as a dispersion spectrometer, and a fluctuation in the frequency of the reference cavity has the same effect as a fluctuation in the klystron frequency. Thus, since P_n' in the stabilizer is equal to $F_A kTB$ (where F_A is the noise factor of the detecting system of the stabilizer), $\overline{\delta f_n^2}$ the mean square value of the imagined fluctuation δf due to noise is given by

$$\overline{\delta f_n^2} = F_A kTB \frac{(\Delta f)^2_{UA}}{P_{CA}} \qquad (6.27)$$

The exact expression for the overall fluctuation δf_{KS} in the frequency of the stabilized klystron is then given by equation (6.28) below, where δf_{KU} represents the fluctuation in the frequency of the unstabilized klystron and δf_R the real fluctuation in the reference frequency.

$$\overline{\delta f_{KS}^2} = \frac{\overline{\delta f_{KU}^2}}{1 + |kG(\omega)|^2} + (\overline{\delta f_R^2} + \overline{\delta f_n^2}) \frac{|kG(\omega)|^2}{1 + |kG(\omega)|^2} \qquad (6.28)$$

The contribution from δf_{KU} here follows from §6.4, provided that $|kG(\omega)|$ is well below the threshold for oscillation. Similarly the remaining contributions are given by $\delta f_{n,R}^2 |kG(\omega)|^2 /[1 + |kG(\omega)|^2]$. All three components are uncorrelated and so are added as the sum of squares.

The extent to which the noise level of the spectrometer is increased is obtained by inserting the above value of δf_{KS}^2 into equation (6.25) for the dispersion spectrometer, and (6.26) for the absorption spectrometer.

The general result is rather complex. However, $|kG(\omega)|$ need not fall below unity until $\omega/2\pi$ reaches a megacycle or two. Thus for all the frequencies of interest the first term in equation (6.28) can be neglected. It is also most unlikely that $\overline{\delta f_R^2}$ will be significant in relation to $\overline{\delta f_n^2}$. Hence the expression obtained by inserting $\overline{\delta f_{KS}^2}$ into equation (6.25) is simplified to give for the dispersion spectrometer

$$\frac{P_n'}{P_n} = \frac{F_A}{F_E} \cdot \frac{P_{CE}}{P_{CA}} \cdot \frac{(\Delta f)_{UA}^2}{(\Delta f)_{UE}^2} \qquad (6.29)$$

There is no reason why F_A and F_E should differ. If the maximum available value of P_{CE} is used, P_{CA}/P_{CE} is equal to the coupling factor (usually 10 dB) of the side-arm coupler that feeds the stabilizer. The ratio $(\Delta f)_{UA}/(\Delta f)_{UE}$, on the other hand, might equal about 1/3, since there is no constraint on the design of the reference cavity other than that it should have the narrowest possible bandwidth, and this makes up for the loss of the coupler. Thus at high values of P_{CE} the value of P_n'/P_n is of the order of unity, so the stabilizer is only marginally adequate, whilst at lower values the fluctuations become insignificant.

§6.6

SUPERHETERODYNE STABILIZER

A good way of avoiding the frequency drift associated with the simple d.c. stabilizer is to convert the detecting system to super-heterodyne operation, preferably using the single-klystron scheme of §5.12. The resulting arrangement is shown in Fig. 6.8, and it will be seen that this is little more than the simple arrangement with a phase-reversing modulator M inserted between the output port of the circulator and the input to the microwave detector. Then, instead of the signal reflected from the cavity causing a change in the d.c. output from the detector, an a.c. signal is produced with a frequency equal to the frequency $f_{i.f.}$ of the oscillator that drives the modulator. This signal is then ampli-fied by an a.c. amplifier A which rejects the d.c. drift in the output from the detector. The a.c. output from the amplifier is then converted to d.c. by a phase-sensitive detector, as in §5.12, and is thereafter used in the same way as the d.c. output from

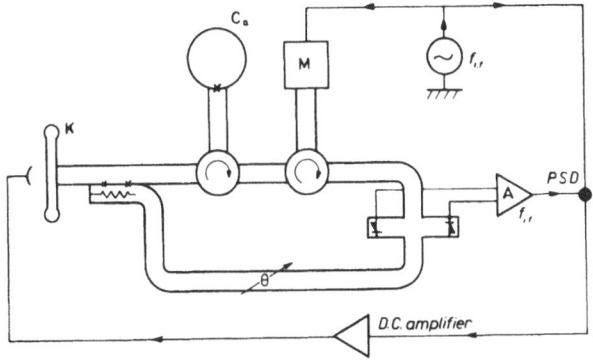

Fig. 6.8. Microwave discriminator stabilizer with superheterodyne detection.

the simple microwave discriminator. Once again it is best to make $f_{i.f.}$ about 30 Mc/s because this value gives the best noise factor. As with the simple discriminator, the noise factor of the detecting system determines the frequency fluctuations obtained with the stabilizer.

The operation of the superheterodyne detector can be very clearly visualized by a slight modification of the diagrams of Fig. 6.2. When the modulator is switched from one state to the other, the phase of the signal reflected from the cavity, which might initially be represented by V_{L1} in Fig. 6.2(a), is reversed and thus adopts the position of V_{L2} in Fig. 6.2(b). The output from the detector, which was originally proportional to $|V_{D1}|$, therefore switches from the value of $|V_{D1}|$ given in Fig. 6.2(c) to that of $|V_{D2}|$ given in Fig. 6.2(d). This switching is periodic so the effect of the signal from the cavity is to superimpose upon the d.c. output from the detector, which is proportional to $|V_{B1}|$, an a.c. signal with a peak-to-peak value equal to the difference between the values of the curves of Figs. 6.2(c) and (d). It is then clear that, for small errors in frequency, the amplitude of the a.c. output is proportional to the magnitude of the error, and the phase of the a.c. output reverses when the sense of the error is reversed. The phase-sensitive detector then converts the a.c. signal to a d.c. signal which is directly proportional to the frequency error, and which can be used in the same way as the output from the microwave detector of the simple discriminator.

§6.7

INCORPORATION OF THE SUPERHETERODYNE STABILIZER
INTO A SPECTROMETER

When the dispersion signal is to be observed, the superheterodyne
stabilizer can be used in exactly the same way as the simple
stabilizer; it can be driven by a small amount of power tapped
from the spectrometer klystron. However, when the absorption
signal is being observed, it is more likely that the klystron will
need to be stabilized relative to the resonant frequency of the
spectrometer cavity. A fairly obvious way of doing this is to
place a phase-reversing modulator at the output of the cavity-
circulator system of the combined simple spectrometer stabilizer

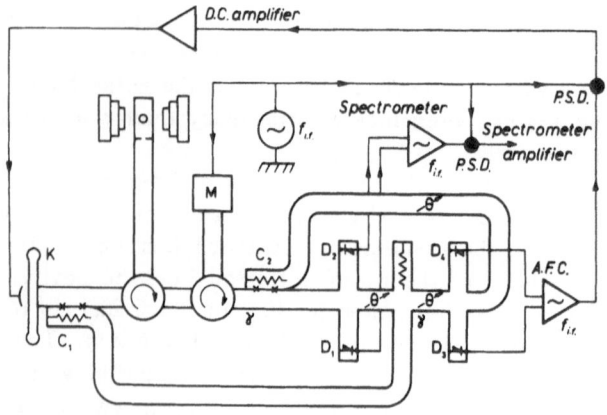

Fig. 6.9. Combined superheterodyne spectrometer and stabilizing system.

of Fig. 6.3. The arrangement is shown in Fig. 6.9, of which the
rest of the circuit is exactly the same as Fig. 6.3, except that
each detector is, of course, followed by an i.f. amplifier and
phase-sensitive detector.

If a simple spectrometer is used, it is still possible to use
superheterodyne stabilization by placing the modulator between
the side-arm of the coupler C_2 and the A.F.C. detector. Alternatively, if, for some unforeseen reason, it is desired to use a
simple stabilizer with a superheterodyne spectrometer— and this

is hardly to be recommended – it is possible to do so by placing the modulator between the main output of C_2 and the spectrometer detector.

§6.8

ASSESSMENT OF SUPERHETERODYNE STABILIZER

We shall now determine the frequency fluctuations associated with the superheterodyne stabilizer, and the remaining, very small component of frequency drift. The method of determination is the same as for the simple stabilizer; the noise and d.c. drift from the detecting system are related to imagined variations in the reference frequency. Then, when the loop gain of the stabilizer is high the imagined variations, plus the real variations in the reference frequency, are transferred to the klystron, whilst when the loop gain is low the original fluctuations in the klystron frequency remain.

The imagined frequency fluctuations are given by equation (6.27), which is the corresponding expression for the simple stabilizer, but F_A, the noise factor of the stabilizer, varies with the frequency rather differently. We shall first verify equation (6.27) for the superheterodyne stabilizer. Suppose initially that the resonant frequency of the reference cavity varies by an amount δf_n. The signal V_L reflected from the cavity then changes by δV_L, which according to equation (6.12) equals

$$\delta V_L = \frac{-jE_g}{2(\Delta f)_{UA}} \delta f_A \qquad (6.30)$$

The phase reversing modulator then periodically reverses the phase of δV_L so that it either adds to or subtracts from the bucking signal represented by V_L. The effect of δV_L on the combined voltage V_D fed to the microwave detector is therefore nearly the same as that of a single-frequency component of amplitude $|\delta V_L|$, but its frequency differs from that of the klystron by the intermediate frequency of the superheterodyne. Since δf_n is a random fluctuation, the resulting noise power is equal to $|\delta V_L|^2/2R_0$, where R_0 is the waveguide impedance. This power represents the noise power $F_A kTB$ of the detecting system, so equation (6.27) follows if P_{CA} is also made equal to $|E_g|^2/8R_0$.

The value of F_A in (6.27) is, however, different from the F_A value for the simple stabilizer. In the simple stabilizer any one component of δf_n of frequency f appears at the output of the d.c. amplifier as a voltage fluctuation of the same frequency. Thus the value of F_A is the noise factor that is measured when the noise is observed through a narrow-band filter that follows the d.c. amplifier and is tuned to the frequency f. This is essentially the condition under which the values in Fig. 2.24(a) were obtained, so the value of F_A for the simple stabilizer increases sharply for the lower frequency components of δf_n. In the superheterodyne system the output from the d.c. amplifier is also directly proportional to the frequency fluctuation δf_n, in spite of the conversion to and from the intermediate frequency. However, the noise centred at a frequency f at the output of the d.c. amplifier originates from the noise emerging from the i.f. amplifier centred on the frequencies $f_{i.f.} \pm f$. This is the i.f. noise from the detector and the primary stages of the i.f. amplifier. Thus the appropriate value of F_A is that corresponding to the intermediate frequency in Fig. 2.24(a), and this holds for all components of δf_n. Since the intermediate frequency is chosen to be about 30 Mc/s for the simple reason that it gives the lowest value of F_A, the value of $\overline{\delta f_n^2}$ given by the superheterodyne stabilizer is therefore always better than that given by the simple discriminator, and at low frequencies is very much better.

Once again it is necessary to determine the frequency response of the control loop of the stabilizer in order to find the frequency at which the loop gain must fall below unity. This presents a slight problem with the superheterodyne stabilizer, because over one part of the loop the control signal is converted to the intermediate frequency and then restored to its own frequency. This section of the loop starts with a variation of, say, δf_K in the klystron frequency, and finishes with a corresponding variation in the output v_D from the phase-sensitive detector. The question is: if the frequency f_k is varied sinusoidally at a frequency $\omega/2\pi$, how does the corresponding variation in v_D vary in amplitude and phase as the value of ω is varied? It will now be shown that a section of the loop responds in the same way as a d.c. amplifier that includes a series of simple, low-pass CR filters, with the cut-off frequency of each filter equal to one

half of the bandwidth of a corresponding stage in the i.f. ampli-
fier. Suppose initially that $\delta f_K = \hat{\delta f_K} \cos \omega t$. Then the output
voltage v_m from the microwave detector is given by

$$v_m = k_m \hat{\delta f_K} \cos \omega t \cdot \cos \omega_{i.f.} t \qquad (6.31)$$

where k_m is the efficiency of the microwave discriminator and
$\omega_{i.f.} = 2\pi f_{i.f.}$. Equation (6.31) can be written as

$$v_m = k_m \frac{\hat{\delta f_K}}{2} [\cos(\omega_{i.f.} + \omega)t + \cos(\omega_{i.f.} - \omega)t] \qquad (6.32)$$

Then, if the i.f. amplifier is assumed to be tuned by simple
resonant circuits, the gain of one stage of the amplifier (the rth)
can be represented by the function $k_{ar} G_{ar}(\omega)$, where k_{ar} is
independent of frequency and

$$G_{ar}(\omega) = \left[1 + \frac{2j\delta\omega}{(\Delta\omega)_{ar}} \right]^{-1} \qquad (6.33)$$

$\delta\omega$ representing the difference between the frequency of the
input to the amplifier and the frequency to which the amplifier
is tuned, and $(\Delta\omega)_{ar}$ the bandwidth of the stage. Now $\delta\omega =
\omega_{i.f.} - (\omega_{i.f.} \pm \omega) = \pm\omega$ for the two single-frequency sideband
components represented in equation (6.32). Thus the upper and
lower sidebands experience phase shifts of $\pm\phi_{ar}$ respectively,
where

$$\phi_{ar} = \tan^{-1} \frac{2\omega}{(\Delta\omega)_{ar}} \qquad (6.34)$$

Each sideband is subject to a reduction of $|G_{ar}(\omega)|$ relative to
the zero-frequency value, where

$$|G_{ar}(\omega)| = \left\{ 1 + \left[\frac{2\omega}{(\Delta\omega)_{ar}} \right]^2 \right\}^{-\frac{1}{2}} \qquad (6.35)$$

Thus the sidebands experience a phase shift and attenuation when
the value of ω approaches one half of the bandwidth of the i.f.
amplifier, which according to §6.4, is the same as that which a
signal passing through a d.c. amplifier experiences when the
frequency of the signal approaches the cut-off frequency of a
d.c. amplifier. Furthermore, the effect of successive stages in
the i.f. amplifier is simply to add to the phase shift and loss of
gain exhibited by one stage. All that remains to be shown in that

the phase shift and attenuation of the sideband are directly transferred to the signal at the output of the phase-sensitive detector.
Now if ϕ_a represents the overall phase shift introduced by the
i.f. amplifier and $|G_a(\omega)|$ the overall reduction gain as ω is
increased, then the signal v_0' at the output of the i.f. amplifier
is given by

$$v_0' = \tfrac{1}{2}\delta f_K k_m k_a |G_a(\omega)| \{\cos[(\omega_{i.f.} + \omega)t - \phi_a]$$

$$+ \cos[(\omega_{i.f.} + \omega)t + \phi_a]\} \qquad (6.36)$$

where k_a is the overall mid-band gain of the amplifier and is
equal to $\overset{n}{\underset{r=1}{P}} k_{ar}$ when there are n i.f. stages. Equation (6.36) can
then be rewritten in the form

$$v_0' = \delta f_K k_m k_a |G_a(\omega)| \cos\omega_{i.f.}t . \cos(\omega t - \phi_a) \qquad (6.37)$$

The phase-sensitive detector then converts this signal to a
sinusoidal output v_D of the form

$$v_D = \delta f_K k_m k_a k_D |G_a(\omega)| \cos(\omega t - \phi_a) \qquad (6.38)$$

where k_D represents the efficiency of the phase-sensitive detector. Thus the phase and amplitude variations of the sidebands,
ϕ_a and $|G_a(\omega)|$ respectively, are, in fact, transferred to the a.c.
error signal at the output of the second detector, with a cut-off
frequency equal to one half of the bandwidth of the a.c. stage.

It is difficult to make the bandwidth $(\Delta f)_{i.f.}$ of the i.f. amplifier more than about 20 per cent of the intermediate frequency
$f_{i.f.}$, which is usually about 30 Mc/s. Thus the cut-off frequency
of the equivalent multi-section low-pass filter $(\Delta f)_{i.f.}/2$ will be
about 3 Mc/s. There is no reason why the bandwidth of the d.c.
section of the loop need be much lower than this value, so it can
be assumed that the stabilizer will operate for frequencies
ranging from 0 to 1 or 2 Mc/s.

We shall now consider the frequency drift obtained with a
superheterodyne frequency stabilizer. One source of drift is the
limiting value of the fluctuation given by equation (6.27). Inserting a value of B equal to the inverse of the time taken to record
a spectrum (say 100 sec) gives an absurdly low value $\delta f_n =$
10^{-6} c/s if the other factors in the equation are typical: $(\Delta f)_{UA} =$
3 Mc/s, $F_A = 10$, $P_{CA} = 1$ mW. Here $\delta f_n = (\overline{\delta f_n^2})^{1/2}$. This factor
can therefore be neglected.

Another cause of drift is d.c. drift in the output from the phase-sensitive detector and the d.c. amplifier. This drift can be referred back to an imaginary drift in the reference frequency, and this is transferred to the klystron. The effects of d.c. drift can clearly be minimized by obtaining as much of the loop gain as is possible from the i.f. amplifier. However, beyond a certain value the noise from the amplifier overloads the phase-sensitive detector. Immediately prior to overload it can be estimated that the drift might typically represent about 1 per cent of the unfiltered noise voltage at the output of the phase-sensitive detector. Thus the drift component of δf_n is equal to 1 per cent of the value of δf_n given by equation (6.27), with B equal to the i.f. bandwidth (about 3 Mc/s). With the above values this gives a drift of 0·5 c/s, again a negligible figure.

Probably the only other factor that can cause drift in a super-heterodyne stabilizer is differential thermal expansion of the waveguide runs. In the composite system of Fig. 6.9, if any part of the waveguide between the points marked γ and γ expands individually, the phase of the bucking signal to the spectrometer will be altered relative to the phase of the signal from the cavity, dispersion will be mixed with absorption, and fluctuations in frequency will be converted to noise. It is therefore a sensible precaution to make these runs as short as possible.

§6.9

POUND STABILIZER

The Pound stabilizer[2] and its near equivalent, the equal-arm stabilizer, are two circuits that have been very widely used in the general field of microwave engineering for stabilizing the klystron frequency. Both are forerunners of the superheterodyne stabilizer described in §6.6, and have a performance that is not very far short of the superheterodyne stabilizer. In fact the only real difference between the Pound and equal-arm stabilizers and the superheterodyne stabilizer lies in the microwave components used; the Pound and equal-arm stabilizers were devised before microwave circulators and P.I.N. modulators were generally available.

Pound and equal-arm stabilizers have been used with E.S.R. spectrometers and so we shall now describe how they work and what modifications have to be made before they can be incorporated into a spectrometer.

The original Pound stabilizer is shown in Fig. 6.10(*a*). The power from the klystron K is first divided between the two side-arms of the magic-tee. One half of the power goes to the detector of the system D_1 and provides the bucking signal, whilst the other goes to the microwave cavity C. Any signal reflected from the cavity is again divided at the magic-tee. One half goes back in the direction of the klystron and is absorbed, preferably by an isolator I which is placed after the klystron. The other half is

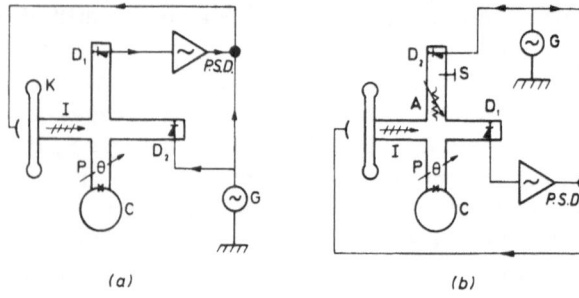

Fig. 6.10. (*a*) Pound stabilizer. (*b*) Equal-arm stabilizer.

fed to the second microwave detector D_2. D_2 performs the same function as the microwave modulator in the superheterodyne stabilizer, or nearly so. The generator G is connected to the output terminals of D_2, and either forward or reverse biases it. D_2 is an ordinary microwave detector and presents a near match to the microwave energy when forward biased and a near open circuit when reverse biased. The reflection coefficient therefore alternates from zero to unity, and so the system works as a chopping modulator. The modulated signal is then returned to the magic-tee and divides once more. Half the power reaches the cavity, which is almost completely matched to the waveguide, and therefore absorbs most of the signal. The other half combines with the bucking signal at the detector D_1. Thus the two signals reaching D_1 are the unmodulated bucking signal and the modulated signal from the cavity. The same situation occurs at the detector of the superheterodyne stabilizer, and therefore the

Pound and the superheterodyne systems are, in principle, identical. It is necessary to adjust the phase of the signal from the cavity relative to the bucking signal so that the detector responds to changes in the reactance of the cavity and not to changes in the effective resistance. The phase shifter P is used for this purpose.

The Pound stabilizer is inferior to the superheterodyne stabilizer in two respects. (a) The power is divided at the magic-tee three times, and this makes for a total loss of 9 dB. (b) A chopping modulator is used in place of a phase-reversing modulator, and this introduces a further loss of at least 6 dB. The overall loss is therefore a little over 15 dB.

The operation of the equal-arm stabilizer is similar to that of the Pound stabilizer, and comparison of Figs. 6.10(a) and (b) reveals that the two arrangements are identical apart from a reversal of the positions of the detector and modulator. The power from the klystron K is first divided at the magic-tee, as with the Pound stabilizer. One half of the power goes to the cavity and the other to the modulating crystal D_2. Both reflected signals are then returned to the magic-tee and are power divided so that half of each reaches the detector D_1. The other two halves are returned to the isolator I and absorbed. Therefore the main difference between the Pound stabilizer and the equal-arm stabilizer is that in the Pound stabilizer the signal from the cavity is modulated, whilst in the equal-arm stabilizer it is the bucking signal that is modulated.

The main advantage of the equal-arm stabilizer over the Pound stabilizer is that, as the name implies, the paths taken by the bucking and cavity signals are of equal length. In the Pound stabilizer the lengths differ, which means that any thermal expansion in the waveguide will alter the phase of the cavity signal relative to the bucking signal. If the cavity is not correctly coupled to the waveguide so that an *in phase* signal is reflected from the cavity, such a phase shift allows the detector to respond to this signal and the klystron frequency is then shifted by the stabilizer to produce a reactive reflection which balances the effect.

Against the above advantage, the equal-arm stabilizer is much more susceptible to changes in the characteristics of the modulator crystal. If the modulator were operated exactly as it is in the Pound stabilizer, that is as a chopping modulator, the bucking signal would be amplitude-modulated to a depth of 100 per cent, even in the absence of any signal reflected from the cavity. It is of course required that only a signal reflected from the cavity should result in amplitude modulation of the signal reaching the detector. The modulator must therefore be operated in a purely phase-reversing mode. This means matching the modulator by means of the slide-screw tuner S so that the magnitude of the reflection coefficient is the same in both forward- and reverse-biased conditions of the diode. Naturally any change in the characteristics of the diode may unbalance the two magnitudes and result in amplitude modulation. The stabilizer will automatically shift the klystron frequency so that a signal, which is strong enough to remove the amplitude modulation, is reflected from the cavity.

The equal-arm stabilizer offers two further minor advantages over the Pound stabilizer. First, in the equal-arm stabilizer the cavity signal is only divided at the magic-tee twice, as opposed to three times in the Pound stabilizer. Second, the level of the bucking drive to the detector D_1 can, in principle, be adjusted to give the optimum noise factor. This can be arranged by inserting an attenuator A before the modulator D_2. It is, however, doubtful whether the phase-modulated bucking signal is, by the time the modulator has been adjusted to give no amplitude modulation, of an amplitude that is above that which gives the optimum noise factor in the first place. In the Pound stabilizer the bucking power is initially one half of the power from the klystron and is probably in excess of the optimum. Unfortunately, however, it is not possible to insert an attenuator to reduce the bucking power without, at the same time, intercepting the cavity signal. In consequence both the Pound stabilizer and the equal-arm stabilizer are a little inferior to the superheterodyne stabilizer, since the noise factor of their detectors is worse than optimum.

In all, the loss of efficiency of the Pound stabilizer and the equal-arm stabilizer relative to the superheterodyne stabilizer is about 20 dB. In other words, the drift and frequency fluctuations are about ten times worse than those obtained with the superheterodyne stabilizer. It is shown in the previous section that the frequency stability afforded by the superheterodyne stabilizer is more than adequate for E.S.R. applications, so the Pound and equal-arm stabilizers can well be used to replace the superheterodyne stabilizer for this purpose.

Pound and equal-arm stabilizers can be used in a dispersion spectrometer with no difficulty, because there the klystron frequency must be stabilized to an external reference cavity. It is therefore sufficient to tap off a modest fraction of the power from the spectrometer klystron, probably by means of a directional coupler, and use the power to drive the stabilizer.

When the E.S.R. absorption signal is to be observed, the klystron should preferably be stabilized relative to the resonant frequency of the spectrometer cavity, and the two stabilizing systems must therefore be modified. One suitable system has been devised by Owston[3] and is shown in Fig. 6.11. Here C

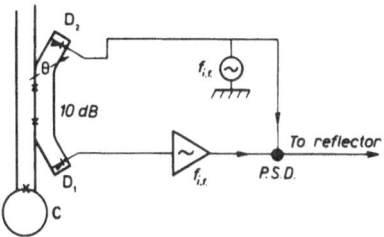

Fig. 6.11. Incorporation of a Pound stabilizer into an absorption spectrometer.

represents the cavity of any normal reflection-cavity spectrometer. The stabilizer operates as a Pound stabilizer in that part of the signal reflected from the cavity passes through the side-arm of the directional coupler to the modulating crystal D_2. The modulated signal is then reflected to the detector crystal D_1, where it combines with the bucking signal which is derived from the coupled fraction of the power incident upon the spectrometer cavity. The remainder of the system is the same as the normal Pound stabilizer. Equal-arm operation can be obtained, if desired, by simply reversing the positions of the detector and modulator crystals. The value of 10 dB is chosen for the coupling factor of the directional coupler because it does not rob the spectrometer of too much power and, at the same time, provides

an adequate signal for the stabilizer. For an incident power of
5 mW on the cavity, the bucking power is optimum (0·5 mW), but
at lower powers the noise factor of the detector of the stabilizer is
degraded. One way of overcoming this problem is to use a
variable isolator adjacent to the cavity to control the cavity

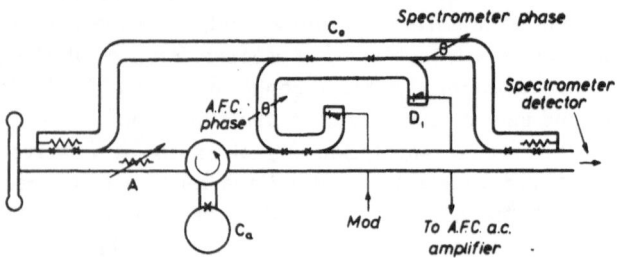

Fig. 6.12. Improved method of incorporating modified Pound
stabilization into an absorption-type E.S.R. spectrometer.

power. Otherwise the arrangement of Fig. 6.12 can be used.
Here the bucking power is derived prior to the attenuator A
which controls the power incident upon the cavity. The coupling
factor of the directional coupler C_0 is chosen to give to the detector D_1 the bucking drive that gives the optimum noise factor.

§6.10

STABILIZATION BY FREQUENCY MODULATION

A very popular method of stabilizing the klystron frequency of an
E.S.R. spectrometer to the resonant frequency of the spectrometer
cavity is shown in Fig. 6.13. The spectrometer S might equally
well be of the transmission, absorption or reflection-cavity type,
but, for the present, we shall assume that a reflection-cavity
system is used. If the absorption part of an E.S.R. signal is to
be observed (and only then is it required to stabilize relative to
the spectrometer cavity), the output voltage from the detector D
of the spectrometer varies with the voltage applied to the klystron
reflector in the manner shown in Fig. 3.8. Moreover, the centre of
the dip corresponds to the reflector voltage for which the klystron
frequency is equal to the resonant frequency of the cavity. If then
a small a.c. voltage from the generator G is applied to the klys-

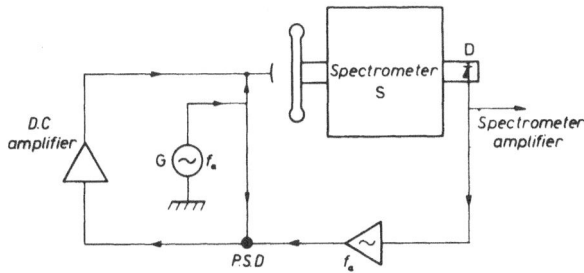

Fig. 6.13. Frequency stabilization by frequency modulation of the klystron.

tron, the klystron is frequency-modulated and the frequency modulation is converted to amplitude modulation, with an efficiency that is proportional to the slope of the curve of Fig.3.8. The slope is zero at the centre of the resonance curve, so the conversion is nil. But if the klystron drifts, amplitude modulation is produced with an amplitude that is porportional to the magnitude of the drift, and a phase that is determined by the sense of the drift. Then, if an amplifier tuned to the modulating frequency f_a is placed after the detector, and the output from the amplifier is coupled to a phase-sensitive detector, the d.c. output from the phase-sensitive detector is proportional to the frequency error, and can be used to control the reflector voltage of the klystron, and thus to stabilize its frequency.

Assessment

This method of stabilization is very good for ensuring that the klystron is tuned exactly to the point of zero response to f.m. noise, because the system does, in fact, set to the position of zero response to frequency modulation. Provided that the cavity is the only frequency-sensitive element within the microwave circuit, this also is the condition for zero response to the dispersion signal. A potential disadvantage of the system is that the frequency modulation of the klystron can degrade the resolution of the spectrometer and must therefore be limited. We shall now consider the extent of the frequency drift, and hence the dispersion suppression factor, S, that is obtained in the light of this restriction.

The method of detection avoids the effects of d.c. drift in the output from the microwave detector, and the only remaining significant causes of drift are the following: drift due to incomplete suppression of the original drift in the klystron frequency, drift due to the lowest significant frequency component of the total effective fluctuation in the reference frequency, and d.c. drift in the output from the phase-sensitive detector and d.c. amplifier. The original drift can be adequately suppressed by making the loop gain of the stabilizer high enough, so only two remaining causes will be considered. Noise from the detecting system of the stabilizer, as in previous cases, can be related to an imagined fluctuation in the resonant frequency of the reference cavity. Then, at low frequencies, where the loop gain of the stabilizer is high, the imagined fluctuations are transferred to the klystron by the action of the stabilizer. We therefore need to find the magnitude of the fluctuation in the resonant frequency of the cavity that will produce a voltage at the input to the spectrometer equal to the noise voltage. Equation (6.44) below gives the mean square value $\overline{\delta f_n^2}$ of this fluctuation δf_n, where F_A is the noise factor of the detecting system, $\hat{f}_d$ the peak frequency deviation of the klystron due to the applied modulation, $(\Delta f)_U$ the unloaded bandwidth of the cavity, P_B the bucking power, and P_C the cavity power. The equation is obtained from Fig. 6.14 which illustrates the effect of frequency modulation. In Fig. 6.14, V_B

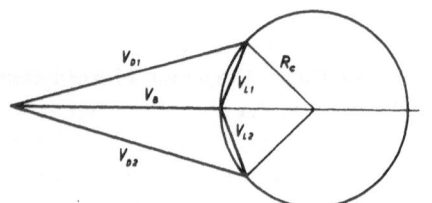

Fig. 6.14. Phasor diagram showing microwave voltages at the two extremes of frequency deviation applied to the klystron in Fig. 6.13.

represents the bucking signal to the detector, and V_{L1} and V_{L2} represent the signal from the cavity when the frequency deviation f_d is at a maximum in the positive and negative directions. The combined signal V_D reaching the detector therefore swings periodically between the extremes of V_{D_1} and V_{D_2} as the klystron is frequency-modulated. If then the mean frequency of the klystron drifts, either $|V_{D_1}|$ grows and $|V_{D_2}|$ shrinks, or vice versa, according to the sense of the drift. In either case $|V_D|$ is amplitude-modulated at the modulating frequency. It can now be imagined that the resulting amplitude modulation is due instead to a microwave signal of amplitude $|V_S| = (|V_{D_1}| - |V_{D_2}|)/2$, and of a frequency that differs from that of the klystron by the modulating frequency f_a. This voltage can then be equated to the noise voltage, and hence, by relating $|V_S|$ to the frequency error δf_n that causes it, the necessary connection between δf_n and the noise voltage can be obtained. It is a fairly simple matter to show from the geometry of Fig. 6.14 that if $|V_B| \gg |V_{L_1}|$ and $|V_{L_2}|$, then

$$|V_{D_{1,2}}| \simeq |V_B| + \frac{1}{2}\left(\frac{|V_{L_{1,2}}|^2}{|V_B|} + \frac{|V_{L_{1,2}}|^2}{R_c}\right) \qquad (6.39)$$

Then, since $|V_S| = (|V_{D_1}| - |V_{D_2}|)/2$ and from Fig. 3.4 $R_c = E_g/4$, where E_g represents the generator voltage associated with the cavity power P_C,

$$|V_S| = \frac{|V_{L_1}|^2}{4|V_B|} + \frac{|V_{L_1}|^2}{|E_g|} - \frac{|V_{L_2}|^2}{4|V_B|} - \frac{|V_{L_2}|^2}{|E_g|} \qquad (6.40)$$

If $\delta f_n = 0$, $|V_S| = 0$, so

$$|V_S| = \frac{d|V_S|}{d(\delta f_n)}\delta f_n \qquad (6.41)$$

Then from equation (6.12), if $\hat{f}_d \ll (\Delta f)_U$,

$$|V_{L_{1,2}}| = \frac{|E_g|}{2(\Delta f)_U}(\hat{f}_d \pm \delta f_n) \qquad (6.42)$$

Inserting this result into equation (6.40), differentiating the result, and inserting the final result into equation (6.41) then gives

$$|V_S| = \left[\frac{E_g}{(\Delta f)_U}\right]^2\left(\frac{1}{4|V_B|} + \frac{1}{|E_g|}\right)\hat{f}_d\,\delta f_n \qquad (6.43)$$

The power associated with $|V_S|$ is equal to $|V_S|^2/2R_o$, where R_o is the waveguide impedance. This power is then put equal to the noise power $F_A kTB$ which, since δf_n is then a fluctuating quantity with a mean square value of $\overline{\delta f_n^2}$, gives

$$\overline{\delta f_n^2} = (\Delta f)_U^2 \frac{F_A kTB}{P_C} \left[\frac{(\Delta f)_U}{\hat{f}_d} \right]^2 \left[\frac{P_B^{\frac{1}{2}}}{P_C^{\frac{1}{2}} + (4P_B)^{\frac{1}{2}}} \right]^2 \qquad (6.44)$$

where $P_B = |V_B|^2/2R_o$ and $P_c = |E_g|^2/8R_o$.

If a balanced detector is used, the previous argument can be modified in much the same way as that of §6.5 was, and gives the result

$$\overline{\delta f_n^2} = \frac{1}{2}(\Delta f)_U^2 \left(\frac{F_A kTB}{P_C} \right) \left[\frac{(\Delta f)_U}{\hat{f}_d} \right]^2 \qquad (6.45)$$

In addition, the plus sign in equation (6.44) is converted to a minus if the phase of the bucking signal happens to be the reverse of that shown in Fig. 6.14.

We shall now insert typical values in equations (6.44) and (6.45). $(\Delta f)_U$ might be 3 Mc/s, $P_C = 1$ mW and $P_B = 0.5$ mW. A popular choice for the modulating frequency is 10 kc/s, so, allowing for inadequacies in the coupling between the detector and the amplifier (it is more important that the coupling from the detector to the spectrometer amplifier should be loss-free, than the coupling from the detector to the A.F.C. amplifier), F_A might be 40 dB. From equation (1.5) a resolution of 0.1 gauss at an operating frequency of 10 kMc/s would require $\hat{f}_d = 300$ kc/s. Then if the spectrum is run in 100 sec, the appropriate value of B for the limiting drift is 0.01 c/s. With these values, $\overline{\delta f_n^2} = 0.036$ for the single-ended detector and 0.09 for the balanced detector. Thus the effective frequency drift is approximately 0.3 c/s in each case. The corresponding value of the dispersion suppression factor S is obtained by inserting these values into equations (6.16) and (6.17) for the single-ended and balanced detectors respectively. These equations indicate that S is in the region of the ratio of the frequency drift to the bandwidth of the cavity so, with the above values, S is more than adequate.

We shall now determine the specific relationship between S and the values given in equations (6.44) and (6.45). This is done by inserting the values of $\overline{\delta f_n^2}$ given by (6.44) and (6.45) into equations (6.16) and (6.17). The result for both the balanced and the unbalanced detector is

$$S = \sqrt{2} \left(\frac{F_A kTB}{P_C} \right)^{\frac{1}{2}} \frac{(\Delta f)_U}{\hat{f}_d} \tag{6.46}$$

It is doubtful whether the resolution required of the spectrometer would ever warrant making $\hat{f}_d$ less than 1/100th of the value quoted. Also, P_C would hardly need to be reduced to more than 1 μW, which is 1/1000th of the value quoted. Under these extreme conditions S rises to about 10^{-3}, which is still more than sufficient. In consequence, drift due to frequency fluctuations need not represent a limitation.

We shall now consider the effects of drift in the output from the phase-sensitive detector and the d.c. amplifier. If $f_a = 10$ kc/s, the bandwidth of the a.c. amplifier would be about 2 kc/s. If, then, the gain were such that the phase-sensitive detector was almost overloaded with noise, the ratio of the total drift would represent about 1 per cent of the unfiltered noise at the output of the phase-sensitive detector. Thus the resulting frequency drift would be equal to 1 per cent of the value of $(\overline{\delta f_n^2})^{\frac{1}{2}}$ given by equation (6.44) or (6.45) (whichever is appropriate) with $B = 2$ kc/s. The results are in each case about 2 c/s which again is negligible.

The general conclusion is that, provided the maximum permissible frequency deviation and the maximum possible gain in the tuned A.F.C. amplifier are used, the present type of stabilizer is perfectly suitable for the application it is required to fulfil and need not degrade the resolution of the spectrometer.

Practical considerations

One or two practical considerations are worth mentioning. One is the choice of the modulating frequency f_a. In order to keep F_A low, it is advisable to make f_a high. However, if f_a becomes comparable with the bandwidth of the cavity $(\Delta f)_U$, and is, at the same time, widely different from the frequency f_m at which the field of the spectrometer magnet is modulated, difficulties can result. All of the foregoing theory assumes that $f_a \ll (\Delta f)_U$, and if this is violated, it is possible that the klystron frequency that gives zero conversion of frequency modulation to amplitude modulation at f_a differs from that which gives zero conversion at f_m. In consequence it is safest to make f_a roughly equal to f_m (but not so close as to be within the band-pass of the spectrometer amplifier), or if f_m is much below the bandwidth of the cavity, to restrict f_a to a value well below the bandwidth of the cavity. Another reason why f_a should be as high as possible is that the bandwidth of the stabilizing loop must be less than f_a. Thus a low value of f_a restricts the loop gain obtainable. Presumably the reason why f_a is commonly dropped to 10 kc/s when f_m has its usual value of 100 kc/s is that it is easier to design a loss-free filter that will separate the two components at the output of the microwave detector if the frequencies differ widely. It seems possible, however, that if f_m and f_a were similar, the two signals might be amplified together by the first one or two stages of the spectrometer amplifier, which could be designed to have a wider bandwidth than the following stages. The following stages could then be designed to reject the A.F.C. signal. When f_a is made much lower than f_m, it is important to ensure that f_m is not a multiple of f_a; otherwise harmonics of f_a interfere with the E.S.R. signal.

Modulation of cavity

When a high resolution is required of the spectrometer, and the frequency deviation $\hat{f}_d$ therefore needs to be small, the loss in stability can be made good by modulating the resonant frequency of the cavity rather than the klystron. Then the deviation can approximate to the bandwidth of the cavity without degrading the resolution at all. One way of modulating the cavity is to vibrate one of its walls, usually by a piezo-electric transducer. This is not a particularly suitable method for an E.S.R. cavity, because the modulating frequency should preferably be quite high (10 kc/s – 1 Mc/s) and the resulting ultrasonic waves could well interfere with the sample. Another method[4] is to couple a normal crystal detector, via a phase-shifter, to a second and somewhat smaller coupling hole in a normal reflection cavity.

When the diode is forward biased, it presents an approximate match to the waveguide, and so a resistive impedance is reflected into the cavity. When the diode is back biased it presents a much higher, but not infinite, resistive impedance. By inserting a suitable length of waveguide between the cavity and the detector, this impedance can be transformed to a complex impedance such that the resistive part of the impedance reflected into the cavity is the same as before, but now a reactance is added. Thus if the diode is periodically switched the damping of the cavity does not alter but its resonant frequency does. Hence the desired frequency modulation is achieved. The effective length of waveguide is adjusted to the required setting by the phase-shifter, and the coupling hole is made small to avoid undue additional loss from being reflected into the cavity. There is little doubt that it would be better to use a PIN diode in place of the conventional diode because the PIN diode switches from a near short circuit to a near open circuit and, when transformed by a suitable length of waveguide, becomes either a negative or positive reactance.

§6.11

STABILIZATION TO A CRYSTAL OSCILLATOR

Where the best possible degree of long-term frequency stability is required of a klystron, quite a good method is to stabilize the klystron frequency relative to a harmonic of the output from a quartz crystal oscillator. The normal arrangement is shown in Fig. 6.15. Here the output from the quartz oscillator O is multiplied in frequency, first by the multi-stage frequency multiplier M, and then by the microwave crystal D_1. The harmonics from D_1 are mixed with a sample of the output from the klystron K in the detector crystal D_2. The output from D_2 is coupled to a tuned amplifier A_1 and then, in one type of arrangement, to a discriminator D. Then, when the difference between the klystron frequency and one of the harmonics is in the region of the frequency to which the amplifier and discriminator are tuned (usually about 30 Mc/s), the d.c. output from the discriminator is proportional to the difference between the difference frequency and the frequency to which the discriminator is tuned. This

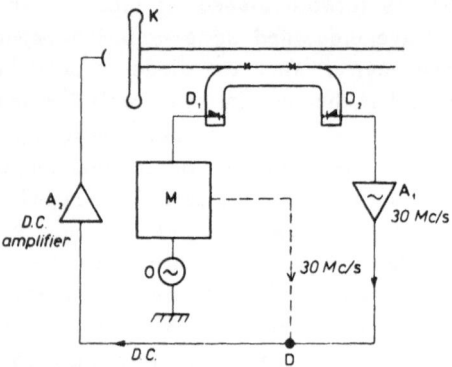

Fig. 6.15. Stabilization of the klystron to the harmonic of a crystal oscillator. K = klystron, M = frequency multiplier, D_1 = multiplier crystal, D_2 = detector crystal, A_1 = i.f. amplifier, D = discriminator or phase-sensitive detector, and A_2 = d.c. amplifier.

voltage can then be fed to the klystron reflector and used to control its frequency. It is clear that the detector D_2 acts as a microwave mixer, with the harmonic acting as the 'signal' and the sample of the klystron output as the 'local-oscillator'. Thus to obtain the best noise factor the level of drive from the klystron to D_2 should be about 0·5 mW [Fig. 2.24(a)]. The value of 30 Mc/s for the amplifier and discriminator frequency is also chosen to give the best noise factor.

The alternative arrangement is to use a phase-sensitive detector (sometimes known in this context as a 'lock-in' or 'phase-lock' detector) in place of the discriminator. The output from a 30 Mc/s oscillator is used to provide the reference for the lock-in detector. Then, with the stabilizing loop open, a beat frequency is produced at the output of the lock-in that is equal to the difference between the difference frequency (between the klystron and the crystal harmonic) and 30 Mc/s. When the loop is closed and the beat signal fed to the klystron, the system 'locks' so that the beat frequency becomes zero. It is not immediately obvious how this happens, but it is possible to see, once it has happened, how the lock is maintained. If the reference and input frequencies to the lock-in detector are the same, and the phase of the input is in quadrature with the reference, then the d.c. output is zero. If then the klystron tends to drift in frequency, the phase of the input signal tends to be changed from its quadrature position, and a d.c. output is developed. If the d.c. output is in the correct sense to oppose the change in frequency, the phase of the signal is restored towards zero and the status quo is preserved. In fact, the sense of the d.c. error signal relative to the frequency change depends on which of the two possible quadrature relationships is held. If this happens to be the wrong one the feedback is positive and the 30 Mc/s signal simply jumps half a cycle to obtain the correct phasing.

The advantage of the lock-in arrangement over the discriminator is that the lock-in detector effectively acts as an integrator in the control loop, and thus gives a zero long-term frequency error. To show this we first imagine that the loop is broken at the output of the lock-in. Suppose then that the frequency of the input signal is exactly equal to the reference frequency, and also that the two have the quadrature phase relationship that is required to give a zero output. Then, if the signal frequency changes instantaneously by a small fraction δf, the phase of the signal begins to vary relative to the reference at the rate of $2\pi\delta f$ radians/sec. The in-phase component of the signal, to which the output from the lock-in responds, is then proportional to $\sin 2\pi\delta f t$, which initially is equal to $2\pi\delta f t$. If δf varies, the corresponding value is $\int 2\pi\delta f t$, and so the system is an integrator. The above argument is only strictly true if the phase error is small compared with $\pi/2$, and this is so when the loop is closed and the stabilizer is in operation. If the potential drift of the klystron becomes so much that the phase error needed to correct it approaches $\pi/2$, the system is in danger of jumping out of lock and becoming unreliable. One therefore always attempts to arrange matters so that the phase error required to correct for the expected variations in the klystron frequency when unstabilized is small. This is ensured by making the gain of the d.c. amplifier A_2 that follows the lock-in detector adequate. The stabilizer is of class 1, since it contains an integrator and, in contrast with the corresponding class 0 system, no single long-time constant is required. Instead, the bandwidth of all sections of the loop should be as wide as possible.

It is a good idea to derive the reference for the lock-in from some section of the multiplier chain, as shown. Then the full long-term stability of the quartz oscillator is transferred to the klystron.

Whilst the long-term stability obtained by using this stabilizer is excellent, the short-term stability may not be. Usually the amplitude of the harmonics from D_1 are rather weak ($<1 \mu W$) so the signal-to-noise ratio of the signal reaching the lock-in detector is poor. The effect of the noise, which is the normal noise from D_2 and the 30 Mc/s amplifier, is to amplitude- and frequency-modulate the 30 Mc/s error signal effectively. Thus, from the point of view of the lock-in detector, the frequency of the harmonic is fluctuating, and this fluctuation is transferred to the klystron in an attempt to cancel the effective fluctuations in the 30 Mc/s error frequency signal.

We shall now determine the effective variation in the harmonic reference due to the noise. The noise power referred to the wave-guide is equal to $FkTB$, where F is the noise factor of the mixer. If P_h is the power of the harmonic, the ratio of the r.m.s. noise voltage $\tilde{v}_n$ to the voltage $|V_h|$ of the harmonic is given by

$$\frac{\tilde{v}_n}{|V_h|} = \left(\frac{FkTB}{P_h}\right)^{1/2} \tag{6.47}$$

Since $\tilde{v}_n \ll |V_h|$, equation (6.47) also represents the effective r.m.s. phase modulation of the harmonic by the noise. Phase modulation always results in frequency modulation, and if f_d is the instantaneous frequency deviation and ϕ_d the corresponding phase deviation, then $\hat{\phi}_d = \hat{f}_d/f_n$, where f_n is the modulating frequency, that is the frequency at which the frequency fluctuation is considered. Thus

$$\tilde{f}_d = f_n \left(\frac{FkTB}{P_h}\right)^{1/2} \tag{6.48}$$

The result emphasizes the general conclusion that the long-term stability is ideal, because when f_n is zero, $\tilde{f}_d$ is zero, but that the short-term stability is poor. The value of $\tilde{f}_d$ for an unstabilized klystron tends to decrease with frequency (Table 6.1) so there must come a value of f_n above which the stabilizer degrades the frequency stability of the klystron rather than improves it. If $F = 10\,\mathrm{dB}$ and $P_h = 1\,\mu\mathrm{W}$, then the value of $\tilde{f}_d$ for a bandwidth B of 1 kc/s becomes $6\cdot5 \times 10^5 f_n$. The corresponding figures shown in Table 6.1 suggest that the transition frequency is about 100 kc/s. Thus if P_h is any lower, or F any higher, it may be necessary, in a 100 kc/s spectrometer, to make sure that the loop gain of the stabilizer is small compared with unity at this frequency. In this way the effective fluctuations of the reference are not transferred to the klystron.

§6.12

VARACTOR FREQUENCY MULTIPLIERS

The method of stabilization described in the previous section will probably soon become outmoded. This is because the varactor frequency multipliers now available give, when driven by a quartz oscillator, a microwave output power that is comparable with that from a klystron (about 100 mW). It is then, possible in principle, to replace the klystron of the spectrometer by one of these sources, and the long-term stability of the quartz oscillator is obtained directly.

The factors that have prevented this technique from being used up to the present are the noise and instability associated with the varactor sources, and the lack of tunability. The varactor chain is capable of adding noise to the microwave output, and also there is a tendency for the multipliers to 'squeg', that is to oscillate in such a way that the amplitude of the output is pulse-modulated at a frequency in the region of a few kc/s. The problem of squegging is now largely overcome, and recent measurements by Frilley and Grandchamp[5] indicate that the problem of noise is now also largely surmounted. The multiplier needs to be tunable for those occasions when it is required to stabilize the frequency of the microwave generator relative to the resonant frequency of the spectrometer cavity. This can be arranged by replacing the quartz oscillator by an electronically tunable oscillator (normally a simple LC oscillator tuned by a varactor) and feeding the d.c. output from the A.F.C. system to the varactor. However, the tuning of the varactor chain is so critical that the slightest change in the oscillator frequency causes the microwave output to fall to an unacceptable degree. Once these problems are fully overcome the klystron will no doubt be replaced, since the whole varactor multiplier and oscillator need only be a few inches in length, and the power supply required is only a simple source of a few volts.

References

1. Chestnut, H., and Mayer, R.W., *Servomechanism and Regulating System Design*, Vol. 1 (Chapman & Hall, 1951).
2. Pound, R.V., *Rev. sci. Instrum.*, 1946, **17**, 490.
3. Owston, C.M., *J. sci. Instrum.*, 1964, **41**, 698.
4. Faulkner, E.A., *J. sci. Instrum.*, 1964, **41**, 347.
5. Frilley, J., and Grandchamp, C., *Design of Low-Noise Solid-State Microwave Sources* (Symposium on Microwave Applications of Semiconductors, London, 1965).

Chapter 7

Low-Noise Microwave Pre-amplifiers

GENERAL CONSIDERATIONS

We shall now consider the advantages likely to be obtained by incorporating into an E. S. R. spectrometer one of the recently developed low-noise amplifiers. In this connection it is probably worth mentioning first that the technology of microwave crystal mixers is still advancing and that the performance of the conventional point-contact silicon crystal rectifier is already being improved upon. Oxley[1] has obtained quite promising results with germanium diodes ($F = 6.5$ dB at 10 kMc/s with an intermediate frequency of 45 Mc/s). The back diode has already been mentioned, with its improved flicker noise characteristics. The newest type of diode is the Shockley barrier diode, which works in much the same way as a point contact diode in that a metal-to-semiconductor junction is used, but the junction is formed by epitaxial growth and has therefore more stable and more controllable characteristics.

The noise factor of a passive mixer is unlikely ever to approach the ideal however, so for the best results some kind of microwave pre-amplifier is possibly an advantage. The most likely alternatives are the travelling-wave tube amplifier, the tunnel diode amplifier, the parametric amplifier,[2] and the microwave maser.[3] The noise factors (4 dB to 6 dB) afforded by the first two are not much better than those for the better types of microwave mixer. Thus, if one is going to complicate matters by adding a pre-amplifier, it is probably better to use either a parametric amplifier or maser; both these give an almost ideal noise figure.

When such a system is used, the noise factor as usually expressed is not a particularly good criterion for describing the

217

noise performance of the system. Generally speaking, the noise factor F of a receiving system is defined as the ratio of the output noise power to that part of the output noise power originating from the source, both powers being measured with the source represented by a resistor generating thermal noise corresponding to a temperature of 300°K (room temperature). We shall now show that pre-amplifiers of the calibre of the maser or parametric amplifier are somewhat wasted on a system with the source at room temperature, and, in such an instance, the concept of 'equivalent amplifier noise temperature' T_a is more useful. In the ordinary way the noise factor F is given by the expression

$$F = \frac{P_n + kT_0B}{kT_0B} \tag{7.1}$$

where kT_0B is the available thermal noise power from the source ($T_0 = 300°K$) and P_n the available noise power from the source that would be needed to account for the noise generated within the amplifier. When the source is at room temperature, the total output noise power P_0 is given by

$$P_0 = G(P_n + kT_0B) \tag{7.2}$$

where G is the power gain of the system. There is therefore little point in using an amplifier with a P_n value that is much less than kT_0B. This point is reflected in the fact that the noise factor tends to a value approaching unity as P_n falls below kT_0B.

Now the equivalent noise temperature T_a of the amplifier is defined by the relation

$$P_n = kT_aB \tag{7.3}$$

and with this definition equations (7.1) and (7.2) become

$$F = 1 + \frac{T_a}{T_0} \tag{7.4}$$

$$P_0 = Gk(T_0 + T_a)B \tag{7.5}$$

Thus for a source at room temperature the above consideration can be expressed by saying that there is no point in using an amplifier with a value of T_a much less than T_0.

The values of T_a afforded by the maser and parametric amplifier are, in fact, for the maser 5°K or less,[3] and for the parametric amplifier $100\,^{\circ}$K for room-temperature operation and $25\,^{\circ}$K if the amplifier is cooled;[4] the maser is inevitably cooled. It is therefore clear that these devices are largely wasted on a system where the source is at room temperature. For the E.S.R. spectrometer the 'source' is the E.S.R. cavity and when the cavity is at room temperature there is certainly no advantage in using anything more refined than the uncooled parametric amplifier. For low-temperature E.S.R. measurements however, when the cavity is cooled, the value of P_0 is given by

$$P_0 = Gk(T_c + T_a)B \qquad (7.6)$$

where T_c is the reduced cavity temperature. It then becomes clear that reductions in T_a are worthwhile up to the point where T_a falls below T_c, and for low-temperature E.S.R. measurements the maser can be used to some advantage. Even for room-temperature measurements this advantage can, in principle, be obtained by cooling the cavity, but not the sample. This could be arranged by inserting a finger dewar into the cavity to keep the sample at room temperature. Such an approach will only work, however, for samples with negligible dielectric loss. When a significant part of the loss within the cavity originates from the sample a corresponding fraction of room-temperature sample noise is engendered.

It should by now be clear that, for room-temperature operation of the cavity, the greatest improvement that can be obtained by adding a low-noise pre-amplifier to a spectrometer system, is equal to the noise factor F_1 of the system before the pre-amplifier is added, but if the cavity temperature T_c is lower than T_0, the potential improvement is $F_1 T_0 / T_c$. For a conventional $100\,\text{kc/s}$ double-modulation spectrometer, $F_1 \simeq 16\,\text{dB}$, and for the super-heterodyne, $F_1 \simeq 10\,\text{dB}$. Thus for room-temperature operation the potential advantages are not great but probably worthwhile, whilst for low-temperature operation of the cavity the potential advantage is considerable.

§7.2

PRACTICAL CONSIDERATIONS

When using a low-noise pre-amplifier it is important that the gain of the amplifier should be sufficiently high for the amplified noise from the source and pre-amplifier to override the noise from the following stages. For this to be so, the pre-amplifier gain must clearly be greater than the ratio between the equivalent noise temperature of the system without the pre-amplifier and the equivalent noise temperature of the pre-amplifier. Expressed slightly differently this means that G_a, the pre-amplifier gain, must be greater than $F_1 T_0 / T_c$, where F_1 is the noise figure of the system without the pre-amplifier. G_a can conveniently be 30 dB but not much higher, which allows for operation at 75°K if $G_a = 3 F_1 T_0 / T_c$ and $F = 16$ dB. The value of 16 dB corresponds to 100 kc/s double-modulation operation of the spectrometer, so if lower-temperature operation is envisaged, or if the modulation frequency has to be lowered, thereby increasing F_1, superheterodyne detection becomes necessary.

When using a reflection-cavity maser or reflection-type parametric amplifier, it is important to remember that these devices are reciprocal in their amplifying properties. This means that noise fed in at the output is amplified just as well as noise fed to the input; indeed the input and output ports are normally one and the same. Thus, unless special precautions are taken, increasing the gain of the pre-amplifier will not ensure that the noise from the source and pre-amplifier will rise above that of the original detecting·system, because noise radiated back from the input of the original system will be amplified as well. The usual way of overcoming this problem is shown in Fig. 7.1. Here the pre-amplifier P is used in conjunction with a four-port circulator C. The signal from the E.S.R. cavity then enters the circulator at

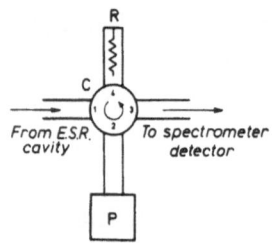

Fig. 7.1 Use of a cavity maser or parametric amplifier in a way that prevents amplification of noise from subsequent stages from occurring.

port 1, is passed out at port 2, is reflected with gain from the pre-amplifier to enter port 2 again and to leave by port 3 in the direction of the detecting system. Noise from the detecting system enters at port 3 and leaves at port 4 to be absorbed by the matched load R. Thermal noise from R, which is usually less than the noise radiated from the detecting system, passes to the E.S.R. cavity, which is usually matched and therefore absorbs the power. In case there is any reflection from the cavity it is usual to cool the load R. The circulator is also cooled to minimize any thermal noise from losses within the circulator.

If a travelling-wave maser or parametric amplifier is adopted the above consideration is of less importance. Normally the travelling-wave systems are largely non-reciprocal in their amplifying characteristics. For the travelling-wave maser in particular a refinement is incorporated that will absorb any wave travelling in the reverse direction.

Another important consideration when using a maser or parametric amplifier is the ease with which either device is saturated. For the maser the input saturation power is 10^{-7} W and for the parametric amplifier the figure is not much higher. It is therefore important, when incorporating either type of device into the spectrometer, to ensure that the bucking signal does not pass through the pre-amplifier. A suitable arrangement for including the pre-amplifier in a simple double-modulation spectrometer is shown in Fig. 7.2. Here the pre-amplifying system S is inserted at the output port of the circulator C normally used in the spectrometer, whilst the bucking signal is combined with the amplified E.S.R. signal at the output of the pre-amplifier. The same system can be used with little modification when it is necessary to use super-heterodyne detection. For double-klystron superheterodyne detection the bucking arm is then broken at the point X, and the local oscillator klystron is inserted at this point. The attenuator A can then be used to provide the optimum local oscillator drive.

It is preferable to use the single-klystron superheterodyne of Fig. 5.10, and here it is simply necessary to insert the microwave pre-amplifier between the output of the E.S.R. cavity circulator and the input of the modulator circulator. Since the modulator effectively splits the signal from the cavity into two sidebands that are separated by twice the modulating frequency $2f_{i.f.}$, it is

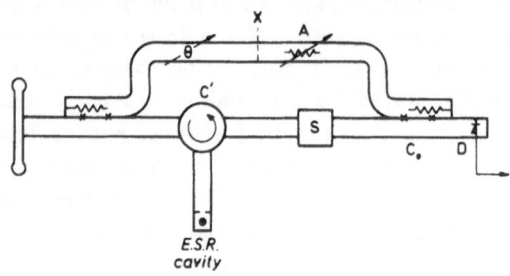

Fig. 7.2 Incorporation of a maser or parametric amplifier system
S into a simple E.S.R. spectrometer.

probably better not to place the pre-amplifier after the modulator;
the bandwidth of the pre-amplifier will probably not be wide
enough to accept both sidebands. In addition, if the amplifier
follows the modulator, noise from the modulator will be amplified
by the amplifier.

If, as is likely, A.F.C. is required, the amplifier should be
placed in the equivalent position in the combined spectrometer
stabilizer of Fig. 6.9. Collins has used[*] a system along these
lines with some success, using however an induction cavity.

When it is possible to use the maximum available power from
the klystron into the E.S.R. cavity, it is very likely that noise
from the klystron will prevent the above improvements from being
obtained. In this instance it becomes necessary to replace the
conventional cavity and circulator by either an induction cavity
(§3.13) or the twin-cavity system of §3.12.

§7.3

COMBINED MASER AND CAVITY

As some interest has been shown in a combined maser-spectrometer system in which the E.S.R. and maser cavity are made one,[5]
it will be valuable to show briefly why the advantages claimed
for this arrangement are somewhat misleading. To this end we
shall consider the equivalent circuit of the combined spectrometer-
maser cavity shown in Fig. 7.3. Here *C* resonates with the induc-

* Private communication from M. J. Collins, N.P.L., Teddington.

tance L in the normal way so that both can be neglected. R_m represents the negative resistance reflected into the circuit by the maser material, R the normal cavity losses, and $\tilde{v}_n$ the noise generator associated with R; we assume the maser to be noiseless. Finally E_s represents the voltage due to the E.S.R. sample.

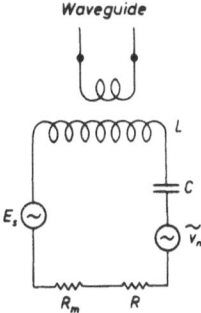

Fig. 7.3 Equivalent circuit of the combined maser-spectrometer cavity.

Prior to excitation of the maser material (and neglecting the absorption in the material) the cavity would, in the ordinary way, be matched to the waveguide. Thus thermal noise power from R and signal power from E_s would be propagated from the cavity to the detecting system, and in each case the power would be the maximum available, that is $E_s^2/8R$ for the signal and kT_cB for the noise, where T_c is the temperature of the cavity. Subsequent amplification by a low-noise amplifier will now increase both noise and signal power equally and will be valuable up to the point where the amplified thermal noise becomes greater than the noise from the remaining part of the spectrometer.

It is then claimed that excitation of the maser material within the cavity, rather than using the subsequent low-noise amplification, increases the intensity of the signal more than it does that of the noise. At first sight it is difficult to see why this should be so. In fact the negative resistance R_m reduces the total effective resistance of the cavity and so increases available noise and signal power from the cavity by the factor $R/(R+R_m)$ – note R_m is negative. In this argument, however, we have neglected to allow for the increased value of E_s. Reduction of the effective cavity losses increases the incident microwave magnetic field in the cavity and E_s is proportional to this field. Thus the available signal power is, in fact, increased more than the noise. Once the maser material is excited it is best to reduce the coupling to the cavity until the cavity is again matched to the waveguide. The somewhat unusual manner of increasing the Q-factor of the cavity makes no difference to the fact that this condition gives the strongest signal from the cavity at the detector. Then all of the increased available signal power is delivered to the detector, whilst the field within the cavity goes up by the same factor as the noise is amplified, i.e. $R/(R + R_m)$. The noise power is therefore increased by $R/(R + R_m)$ and the signal by $R^2/(R + R_m)^2$, so the voltage signal-to-noise ratio is increased by the factor $R^{1/2}/(R + R_m)^{1/2}$.

The improvement is seen to be of little value when viewed in this way because it is achieved by an increase in the microwave field in the cavity. This increase is not usable for saturable samples, whilst for unsaturable samples it could more easily be obtained by a simple increase in the klystron power. In reality the system is completely unworkable because the order of microwave field value that one normally has in an E.S.R. cavity would utterly saturate the maser material from the outset.

References

1. Oxley, T., *Recent Advances in Germanium Mixer Diodes* (Symposium on Microwave Applications of Semiconductors, London, 1965).
2. Sims, G.D., and Stephenson, I.M., *Microwave Tubes and Semiconductor Devices* (Blackie, 1963).
3. Siegman, A.E., *Microwave Solid-State Masers* (McGraw Hill, 1964).
4. Uenohara, M., and Sharpless, W.M., *Proc. I.R.E.*, 1959, **44**, 1183.
5. Townes, C.H., *Phys. Rev. Letters*, 1960, **5**, 428.

Chapter 8

The Spectrometer Magnet System

We shall now discuss the magnet that provides the main magnetic field required by the spectrometer and also the power supply that drives the magnet. The general topic of magnets and magnet power supplies is a subject in its own right and many books[1] are already devoted to the subject.

The requirements of the magnet are (a) that the field should be high enough to satisfy the condition for resonance that is described by equation (1.5), and (b) that the field should be sufficiently homogenious and stable. A high degree of homogeneity is required so that large samples can be accommodated without inhomogenious broadening of the resonance line, and the field must be stable to avoid distortion of the X-axis of the spectrum when it is being traced. The field must also be variable in order to allow a spectrum to be traced, and must be capable of adjustment over a fairly wide range of values to allow for the widely differing g-values that are met in practice. This last requirement can be waived to some extent if the spectrometer is only required for studying free radicals, because for these the g-value is always very close to the free spin value of 2. Indeed, for free-radical studies it is possible to use a permanent magnet with the field adjusted to the value corresponding to $g = 2$, and to vary the field through the resonance condition by passing a current through a suitable pair of sweep coils that are mounted on the pole pieces. In any other instance an electro-magnet is almost obligatory. We shall now consider the above considerations in slightly more detail.

§8.1

HOMOGENEITY

The fundamental factor that determines the homogeneity of the magnet is the area and separation of the pole faces. Generally speaking, the larger the poles and the closer the spacing the better the homogeneity. The alignment of the pole faces is of course also important; a wedge-shaped gap will clearly distort the field. Again micro-inhomogeneities in the make-up of the iron can, in principle, be troublesome, and the pole faces must

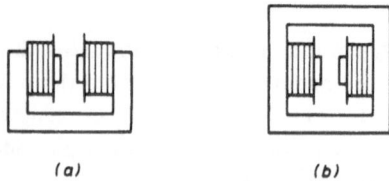

(a)   *(b)*

Fig. 8.1. (a) Single-yoke electromagnet. (b) Double-yoke electromagnet.

be perfectly flat. There is also some advantage in using the double-path type of yoke shown in Fig. 8.1(b) in preference to the single-path system of Fig. 8.1(a), for not only is the double-path arrangement more symmetrical, but it also prevents the gap from tapering owing to bending of the yoke when the field is applied.

If all of these basic considerations are taken care of, the only remaining measure that can improve the homogeneity is to use shims, which fall into two types, ring and current shims. Ring shims are shown in Fig. 8.2(c). In the absence of shims the field tends to decrease with the distance from the axis of the poles, as in Fig. 8.2(a). The effect of the shims is to concentrate the field around the circumference of the poles and to cancel, to some extent, the decay. This is illustrated in Fig. 8.2(b) where the shims are shown on infinite pole faces, and in Fig. 8.2(c) where they are shown on finite faces. It is possible to dimension the shims so that the square and quartic terms in the expansion for the magnetic field (equation 4.27) are cancelled.[2] Quite often provision is made for the shims to be adjusted so that this condition can be obtained.

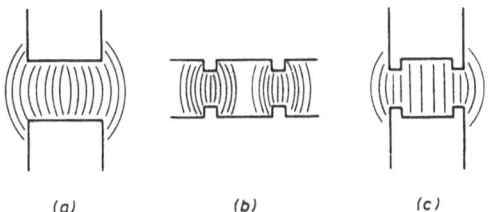

(a)   *(b)*   *(c)*

Fig. 8.2. Effect of ring shims. (*a*) Inhomogeneity without shims, (*b*) effect of shims on infinite pole faces, and (*c*) homogenizing effect of shims.

Current shims are somewhat more complex.[3] A set of current shims is shown in Fig. 8.3 and, in fact, consists of a set of flat coils pasted on the poles of the magnet. These are so arranged

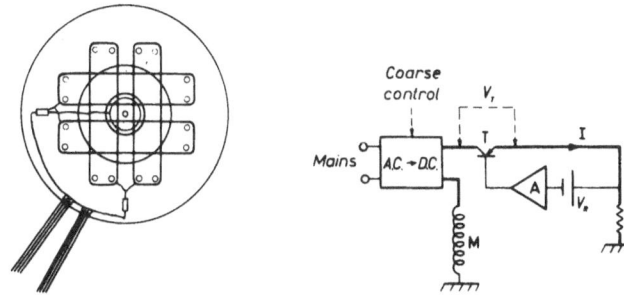

Fig. 8.3. Current shims using a combination of rectangular and circular current loops.

Fig. 8.4. Current stabilizer for a magnet.

that, by suitably adjusting the d.c. currents through the various coils, higher order terms in the inhomogeneity can be cancelled.

§8.2

CURRENT STABILIZATION

In order to obtain a sufficiently stable field some form of stabilizer is required. Magnetic resonance linewidths can be as low as a few parts in 10^5 of the field itself, and an unstabilized supply derived from the electrical mains will clearly not be adequate. Methods of stabilization fall into two classes, field stabilizers and current stabilizers.

Current stabilization is the simplest to arrange and will be considered first. The basic arrangement is shown in Fig. 8.4. Here the current through the magnet M is passed through a standard resistor R to develop a voltage proportional to the current I. This voltage and a reference voltage V_R are then compared by a difference amplifier A, and the amplified error signal is used to control the resistance of a bank of transistors T placed in series with the magnet and the d.c. magnet supply. One of the practical problems with this system is to ensure that too great a power is not dissipated in the series transistors. This demands that the voltage of the d.c. supply should be kept within fairly broad limits. Thus often the voltage V_T across the series transistors is compared with a second reference and used to control, to a certain extent, the voltage of the d.c. source. Two methods are available for doing this. One is to use a normal type of electronic power supply, but with silicon-controlled rectifiers. The rough error signal is then used to control the firing interval of the rectifiers. Another method is to use an a.c. to d.c. rotary converter to supply the primary direct current. Then, by suitable arrangement of two field windings, it is possible to obtain a very close control over the output voltage by altering the current in one of the windings. This current is then determined by the rough error signal. The advantage of the rotary converter over the electronic supply is that much higher ripple frequencies are obtainable, and these are easier to smooth.

Whilst current stabilization represents a considerable advantage over having no stabilization at all, it is lacking in one or two respects. The ability to reset the field to any one value is limited because of hysteresis in the magnet. For the same reason, a linear current sweep may not give a linear field sweep. In addition, current stabilization does not allow for changes in the dimensions of the magnet caused by thermal expansion, or for fluctuations due to external magnetic disturbances. Thus it is better if the field, rather than the current, can be stabilized. The main methods of field stabilization are derivative field stabilization, Hall-effect stabilization, and proton resonance stabilization.

§8.3

DERIVATIVE FIELD STABILIZATION

If a coil is placed in the field of a magnet and the field varies, there develops within the coil an e.m.f. that is proportional to the rate of change of field. Then if the output from the coil is coupled to the d.c. amplifier that controls the series transistors in Fig. 8.4, it will maintain a zero rate of change of field. Furthermore, if a d.c. reference supply is connected in series with the coil, the stabilizing system will ensure that the rate of change of field is exactly sufficient to produce a voltage across the coil that is equal and opposite to the reference voltage. Thus a constant rate of change of field is obtained, which is exactly what is required to sweep out the E.S.R. spectrum. The disadvantage of the derivative field stabilizing system is that any d.c. error voltage or drift in the d.c. amplifier will produce a field error which is the time integral of that voltage. Thus, the greater the time of sweep the less effective the system. However, it is quite adequate in practice for most applications.

§8.4

HALL-EFFECT STABILIZATION

The Hall-effect stabilizer stabilizes the magnetic field directly. If a semiconductor H carrying a current i is placed in a magnetic field B, as shown in Fig. 8.5, the effect of the field is to exert a force on the current carriers, tending to make them move at right angles to their direction of motion and to the field. Hence a charge is built up on one side of the semiconductor and therefore a voltage appears across the faces. This voltage is strictly

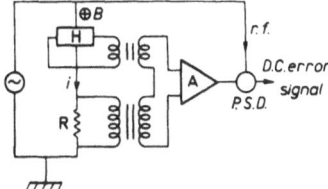

Fig. 8.5. Hall-effect field stabilizer.

proportional to field and can be compared with a reference voltage and used to develop an error signal in the normal way. The Hall probe is the basis of the Varian Fieldial regulating system.

In order to avoid errors due to drift in the d.c. amplifier that would otherwise follow the probe, it is usual to pass an a.c., rather than a d.c., driving current through the probe. The output signal is then an alternating voltage and an a.c. amplifier can be used. Then, in order to convert the a.c. signal back to d.c., it is necessary to feed the output from the amplifier A to a phase-sensitive detector.

When a.c. drive is used, the reference signal must also be an alternating voltage. This is conveniently arranged by passing the driving current through a resistor R, and connecting the resulting voltage in series with the output from the probe. In this way the drive current does not need to be stabilized.

A major problem with the Hall probe is that the output voltage is inversely proportional to the density n of the charge carriers within the semiconductor, and this quantity is temperature dependent. The temperature dependence stems from the fraction of the carriers that are generated thermally (the 'intrinsic' carriers). Thus the doping must be sufficiently high for the number of impurity carriers to be well in excess of the intrinsic carriers. This makes for a high value of n, and so reduces the voltage developed. In a d.c. system this reduction increases errors due to drift voltage and, in an a.c. system, errors due to noise. It is a good idea to keep the probe in close thermal contact with the magnet, because this reduces at least short-term temperature variations.

§8.5

PROTON RESONANCE STABILIZATION

The standard high-precision method of measuring a magnetic field is to use the effect of N.M.R. Basically the proton resonance meter, in conjunction with the magnetic field it is measuring, is an elementary N.M.R. spectrometer. The hydrogen nucleus is generally used and, because the relation between the field and frequency (4258 c/s per gauss) is well known, and of course constant, the field can be measured by measuring the frequency.

A typical proton resonance meter[4] is shown in Fig. 8.6. This is basically a two-stage oscillator with the second stage acting as a limiter. The output is therefore constant and is fed back to the sample coil via a very high impedance which therefore approximates to a current generator. When N.M.R. occurs the Q of the tuned circuit is damped, its impedance drops, and the voltage fed to the broad-band amplifier also drops. This drop is then detected by the grid of the limiter.

The proton resonance meter can also be used to stabilize the field. Generally for the purposes of stabilization one requires a method of measurement that will produce a d.c. voltage proportional to the error in the field. This is then amplified and fed to the control transistors. The direct output from the proton resonance meter is however a steady d.c. voltage that dips as the resonance is traversed. The answer to this problem is to arrange for some form of double-modulation detection, so that the derivative of the absorption signal is produced; this is an S curve and satisfies the requirement at least in the centre of the

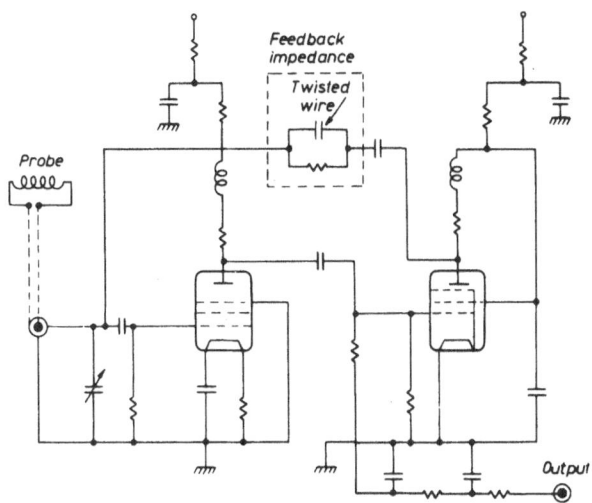

Fig. 8.6. Proton resonance meter circuit. (After the design used by Newport Instruments Ltd.)

resonance. In practice, either the frequency of the oscillator is slightly frequency-modulated, or else the field in the region of — and only in the region of — the N.M.R. sample is. The resulting a.c. component at the output of the proton resonance meter is then amplified by a suitable amplifier and passed to a phase-sensitive detector in the usual way. The output from the phase-sensitive detector then provides the error signal.

<h2 style="text-align:center">§8.6</h2>

<h3 style="text-align:center">FIELD SWEEP</h3>

With all systems of stabilization it is necessary to have some method of sweeping the field of the magnet to trace out the E.S.R. spectrum. With current and Hall-effect stabilization this is simple because a voltage is developed that is proportional either to the current or to the field. This voltage is compared with a standard voltage and the standard voltage is adjusted to give the required field. Thus it is simply necessary to add a small sweep voltage to the reference. Then the field sweeps just fast enough to balance the sweep in the reference voltage.

Where stabilization is to the proton resonance meter matters are not quite so simple. It is the frequency of the proton resonance meter that must be both stable and variable. This is perhaps why the method is not much in favour for stabilization. It is more common to use one of the other types of stabilization, and to use the proton resonance meter to measure the field at the centre of the recorded spectrum.

In one very elegant system[5] a series of quartz crystals is arranged on a turret to control, in turn, the frequency of the proton resonance. As the field sweeps past the value at which the first crystal allows the proton resonance to occur, a point is marked on the E.S.R. spectrum, and then the next crystal is switched in automatically. This process continues and so a whole series of field markers are placed on the spectrum.

Another method (Newport Instruments Ltd) is to start with derivative field stabilization and a steady reference voltage so that the rate of change of field is constant. The proton resonance meter is then locked to any one of a series of harmonics that is developed by a crystal oscillator and frequency multiplier. The frequency of the crystal is chosen so that each harmonic corresponds to an integral field value (via the proton resonance). Thus by locking the oscillator to one of the harmonics one is selecting an integral field value at which the proton resonance will be observed. As the field sweeps past this value the chart recorder is started, so the starting point of the spectrum is accurately known. This is a more flexible system than the previous one, because there the number of quartz crystals is obviously limited. The crystal-turret method is most suitable for free-radical studies, where all the resonances come at about the same field value. Then all the crystal frequencies can be grouped around this point. The harmonic system allows the marker point to correspond to any integral field value, but only one marker per spectrum is obtained.

§ 8.7

FREQUENCY RESPONSE OF CONTROL LOOP

In designing a magnet control system it is important to remember that there is usually a considerable lag between a change in the current fed to the base of the control transistor and the resulting change in magnet current or field. This lag can well provide the one large time constant necessary in a type 0 servo system, but, in any case, it must be taken into account in synthesizing the required overall open-loop frequency response.

References
1. de Klerk, D., *The Construction of High Field Electromagnets* (Newport Instruments, 1965).
2. Bjorken, J.D., and Bitter, F., *Rev. sci. Instrum.*, 1956, **27**, 1005.
3. Anderson, W.A., *Rev. sci. Instrum.*, 1960, **32**, 241.
4. Robinson, F.N.H., *J. sci. Instrum*, 1959, **36**, 481.
5. Horsfield, A., Morton, J.R., and Moss, D.G., *An Automatic Method of Magnetic Field Calibration using Proton Resonance* (National Physical Laboratory Report BP 25/61, 1961).

Chapter 9

Electronic Circuitry

It will be useful in concluding to say something about the electronic circuits included in the design of an E.S.R. Spectrometer. In this chapter in particular the general policy of assuming the background of a graduate in electronic engineering will be followed. There is no longer any need for the circuitry to be designed by the scientist engaged in electron-resonance studies. Most spectrometers in use today are built commercially and it is therefore the professional engineer to whom the chapter must be addressed.

In principle, the circuits used in present-day spectrometers do not differ greatly from those of five or even ten years ago, but the change-over from electron-tube to solid-state circuits has brought some new problems and opportunities. It is also probably true that the tendency towards using integrated microcircuits* in the design of electronic equipment has not been exploited as much as it might in the design of spectrometers. Thus in order to make the treatment as up-to-date as possible and to give this new tendency something of an airing. I have, immediately prior to writing this chapter, designed, built and tested a complete spectrometer. The chapter therefore gives a stage-by-stage description of the various sections of the system and concludes with a description of its overall specifications.

§9.1

BLOCK DIAGRAM

The spectrometer that I constructed is designed along the lines most commonly followed by commercial manufacturers. It has

---

* For an alternative viewpoint see Faulkner, E.A., Grimbleby, J.B., and Wilmshurst, T.H., *J. Sci. Instrum.*, 1967, **44**, 882 double-modulation detection operating at a frequency of 100 kc/s and the technique of frequency modulating the klystron stabilizes the klystron, frequency to the resonant frequency of the spectrometer cavity. The modulating frequency has the usual value of 10 kc/s. The principle of double modulation has been described in §2.3 and the method of A.F.C. in §6.10. The way in which the two systems are combined is shown in Fig.9.1. It should be noted that a wide-band amplifier follows the microwave detector and that this amplifier accepts both the A.F.C. and the E.S.R. signal. This is a very convenient arrangement since it avoids the need for two separate detectors. The only hazard with such a scheme is that the measures adopted to provide the wide bandwidth may degrade the noise figure of the amplifier, but this can be avoided.

§9.2

100 KC/S AMPLIFIER

The complete system for amplifying the 100 kc/s spectrometer signal is shown in detail in Fig. 9.2. The requirements of the amplifier are a good overall noise figure, linearity, good gain stability, good phase stability, suitable bandwidth and frequency response, and adequate and controllable gain. The onus of providing a good noise figure falls largely upon the first stage, that is the wide-band pre-amplifier. Provided that the attenuation placed between the pre-amplifier and the main amplifier is not too high, the greater part of the total noise originates from the first stage in the ordinary way. From this point of view it is therefore advisable to have the attenuator that controls the gain of the system as far forward in the chain of amplifiers as possible. However, if the attenuator is placed too close to the output, the stronger signals tend to overload the earlier stages. In fact, in order to preserve linearity it is best to place the attenuator as close to the input of the system as possible, indeed to place it immediately after the microwave crystal. To satisfy both considerations the best arrangement is to insert an attenuator after the first stage and to allow an attenuation that is variable from zero to a value not more than the gain of the first stage. Any further attenuation is then inserted in a similar fashion after subsequent stages.

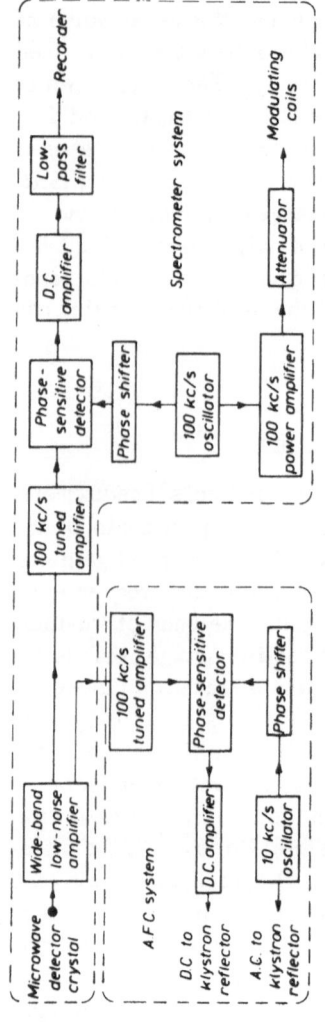

Fig. 9.1. Simplified block model of the electronic section of a 100 kc/s double-modulation spectrometer with A.F.C.

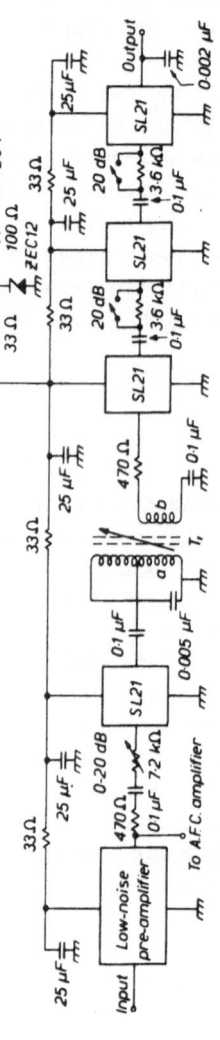

Fig. 9.2. Block diagram of a 100 kc/s amplifier, showing tuning, interstage coupling, attenuating and power-supply decoupling arrangements.

T_1: $a = 73$ turns of 36 SWG tapped 16 turns from earth; $b = 16$ turns of 36 SWG. Both are on Vinkor LA 2704 core.

Gain stability is a most important consideration. If the spectrometer is to be used for quantitative assessment of radical concentration, it is essential that the gain of the spectrometer should not change between running a test sample and running a standard sample. Phase stability is important for the same reason, because the cosine law of the phase-sensitive detector converts phase changes to amplitude changes. In practice, phase drift is not likely to be a major problem, but if poor-quality tuned circuits are used in the 100 kc/s amplifier the effect can be significant. A phase shift occurs if one of the tuned circuits drifts from resonance.

In order to obtain a high degree of gain stability, and also to simplify the problem of design, Plessey type SL21 integrated-circuit shunt-series feedback coupled pairs are used to make up the bulk of the amplifying system. The circuit diagram of one of these elements (see Fig. 9.3) comprises two directly coupled common emitter stages on one chip of silicon substrate. The stages are connected in cascade to give a current gain of 50 dB without feedback, and then one twentieth of the current through the output transistor is fed, via the network in the emitter circuit, back to the base of the input transistor. The gain with feedback then becomes 26 dB and the gain stability is thereby improved by a factor of 24 dB (over ten times) over the value without feedback. The circuit requires a current input and has an output impedance of 400 Ω determined by the resistor in the collector of the output transistor. Thus the stages can be simply cascaded with suitable d.c. blocking to give any gain required with a high degree of gain stability. Furthermore, attenuation is simply arranged by switching a resistor between the two stages, as shown between the latter stages of Fig. 9.2. Decoupling is conventional, using a 12 V Zener diode to define the supply voltage and RC networks to give the required a.c. decoupling.

The operating point of the feedback coupled pair is stabilized by virtue of the fact that the current through a silicon transistor changes abruptly when the base voltage rises above 600 mV. Below this figure the collector current is very low and above it the collector current rises rapidly. In consequence the current through the second transistor is stabilized to give an emitter voltage of a little over 1·2 V, and the base voltage, together with

the collector voltage of the first stage, is kept at rather more than
1·8 V. These figures allow a peak-to-peak output voltage at the
collector of about 2·5 V, with almost complete linearity.

The additional 470 Ω resistor inserted between the pre-amplifier
and the first SL21 stage in Fig. 9.2 is included to prevent the
gain of the 10 kc/s system from being altered too drastically when
the setting of the 7·2 Ω attenuator is varied. The input impedance
of the SL21 is extremely low, being essentially a current input,
so, without the 470 Ω resistor, the 10 kc/s signal would be almost
completely lost in the position of zero attenuation. As it is, the
output impedance of the pre-amplifier is 400 Ω, so the 10 kc/s
gain suffers a variation of about 6 dB as the attenuator is varied.

The amplifier has to be frequency-selective to some extent to
avoid the system having spurious responses at harmonics of the
operating frequency of 100 kc/s. This point has been discussed
in detail in §2.3, where it was shown that the amplifier must reject all harmonics of 100 kc/s and in particular the odd harmonics.
In contrast, the bandwidth needs to be as high as possible for
applications where the response time of the spectrometer is required to be rapid. In the arrangement of Fig. 9.2, the desired
response is obtained by one tuned circuit coupled by empirical
adjustment to give a bandwidth of 10 kc/s. Only one tuned stage
is used to minimize the potential phase drift. The 0·002 μF output decoupling capacitor is introduced to reduce the high-frequency
gain of the final three stages. Without this capacitor there is a
tendency towards regeneration at a frequency of several megacycles. The capacitor does not influence the gain at 100 kc/s.

§9.3

LOW-NOISE PRE-AMPLIFIER

The circuit diagram of the low-noise wide-band pre-amplifier is
shown in Fig. 9.4. It is not really advisable to use the integrated
circuit feedback pair in this position, because the feedback pair
is not designed with the noise factor in mind as the primary consideration. Feedback is necessary in order to give the required
degree of gain stability, but special precautions have to be taken
to ensure that the application of feedback does not degrade the
noise factor. In addition, the value of collector current that gives
the optimum noise figure for a transistor is somewhat lower

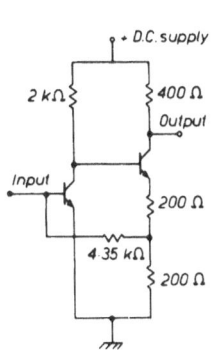

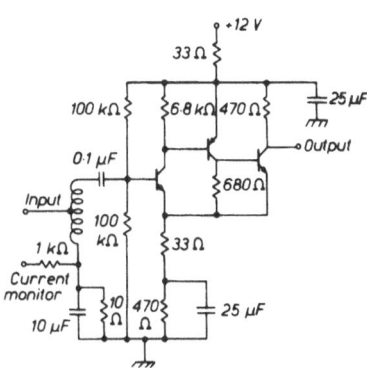

BC114   V405A   C111E

Fig. 9.3. Circuit diagram of a Plessey SL21 shunt-series feedback pair.

Fig. 9.4. Circuit diagram of a low-noise wide-band pre-amplifier. Coil details: 270 turns of 36 SWG tapped 54 turns from end, on GEC152C former.

than that which gives maximum gain and bandwidth.

It is a well-established fact that, in principle, the application of feedback, whether positive or negative, does not alter the noise factor of an amplifier. It is, however, possible for the thermal noise associated with the feedback element to add to the overall noise level and cause a deterioration. For example, in the SL21 feedback pair, noise from the $4·35$ kΩ feedback resistor would probably play a significant part in determining the overall noise level.

In Fig. 9.4 transimpedance feedback is used and is arranged by passing the current from the last two transistors through a 33 Ω feedback resistor connected in the emitter circuit of the first transistor. The value of the feedback resistor is made deliberately low, so that it is small compared with the intrinsic emitter resistance of the first stage. Then the thermal noise from the feedback resistor is small compared with the noise associated with the emitter resistance of the transistor. The first transistor operates at a collector current of $0·1$ mA and so the emitter resistance ($ = 25$ mV$/I_e$) is 250 Ω and therefore large compared with 33 Ω.

The need to keep the feedback resistor low in comparison with the emitter resistor is something of an embarrassment, because it means that the feedback loop gain suffers a loss in proportion to the ratio of the two resistances at that point. In addition, the current gain from emitter to collector of the first stage is only of the order of unity, so in order to obtain a high enough overall loop gain to provide the required degree of gain stability, it is necessary to use three transistors rather than two. Assuming that the second two each have a current gain of 30 and that the loss due to the low feedback resistor is one-tenth, then the loop gain, and therefore the gain stabilization factor, is 10, which is the same value as is afforded by the SL21.

The BC114 is selected as being a transistor specially designed for low-noise operation over the frequency range in question, 10 kc/s to 100 kc/s. The complete circuit comprises three, complementary, grounded emitter stages. The first stage is operated at a collector current of 0·1 mA, this being the value that gives the optimum noise factor, and the second and third at 1 and 10 mA respectively. The current gain of each stage is about 30, so collector load resistors are added to accommodate the excess collector current. Both the 6·8 kΩ and the 680 Ω resistors are chosen to give the normal operating base voltage of 600 mV at the collector currents of 0·1 and 1 mA. The input bias potentiometer and common emitter bias resistor are chosen to give a d.c. voltage of 4 V at the emitter of the first transistor. The 470Ω load resistor in the collector of the final stage is chosen to give a convenient voltage drop and an output impedance comparable with that of the feedback pairs in the remainder of the system. The 10 Ω resistor in the input circuit is included to allow monitoring of the crystal current by a meter or by an oscilloscope. The 10 μF shunting capacitor affords an effective short circuit at 100 kc/s, but allows the klystron mode to be displayed on an oscilloscope with a 50 c/s sweep without too much distortion.

The input transformer is designed to give the optimum 'noise matching' between the output impedance of the microwave detector and the input of the pre-amplifier. It is important to appreciate that, in any low-noise system, the matching condition between generator and amplifier that gives the best noise figure is not the same as that which gives the best power transfer. When the noise figure is potentially poor the two conditions are the same, but otherwise the optimum noise figure is obtained at a degree of over-coupling from the point of power transfer. The optimum noise figure obtained with the circuit of Fig. 9.4 was 1 dB.

We shall now consider rather more generally the factors that govern the low-noise operation of transistors. The simple theory for the noise performance of a transistor amplifier in the absence of flicker noise has been worked out by Nielsen.[1] From this treatment the noise figure F_{opt} for noise matching is given by the expression

$$F_{opt} = 1 + \frac{2r_x}{a_0 r_e}\left[\left(1 + \frac{r_e}{r_x}\right)\left(1 - a_0\right) + \frac{f^2}{f_a^2}\right] \tag{9.1}$$

where r_x is the ohmic or spreading resistance of the base region of the transistor, a_0 the zero-frequency emitter-to-collector current gain, r_e the intrinsic emitter resistance of the emitter and f_a the alpha cut-off frequency of the transistor, i.e. approximately the frequency at which the value of β falls to unity. Here β is the base-to-collector current gain and is given by

$$\beta = \frac{a}{1 - a} \tag{9.2}$$

where a is the emitter-to-collector current gain. The collector current I_c influences F_{opt} in that r_e is given by the expression

$$r_e = \frac{kT}{q_E I_e} \tag{9.3}$$

where I_e is the emitter current $(\simeq I_c)$ and q_E the charge on an electron. For values of $f \ll f_a$, it is then clear that r_e should be as high as possible, but there is little improvement beyond the point where r_e rises above r_x. In practice, matters are complicated by the fact that β is not independent of I_c but falls with it rather slowly. Were this not so, the value of F_{opt} would approach a final value of $1 + 2/\beta$ with decreasing I_c. Because of the variation of β however, the noise factor reaches a minimum at some value of I_c less than the value at which $r_e = r_x$, and the optimum value of noise figure is rather more than the value of $1 + 2/\beta$.

In general we can therefore say that the broad requirements of a low-noise transistor are a high value of β and one that does not fall too rapidly with I_c, and a low value of r_x. If r_x is high,

I_c will have to be reduced a great deal to make r_e approach r_x, and then β is likely to suffer. A low value of r_x and a high value of β are conflicting requirements, because for β to be high the base must be thin and r_x is thereby increased. However, since β is inversely proportional to the square of the width W of the base, whilst r_x is directly proportional to W, it is more profitable to make the base as thin as possible. Thus a low-noise transistor will typically have a high initial value of β that will in fact be reduced to a fairly normal sort of value by reducing the collector current.

At higher frequencies matters are rather different. Because of the f/f_α term in equation (9.1), the noise figure becomes worse as the operating frequency f increases. If $r_e \gg r_x$, as it will probably be, the upper noise corner frequency, that is, the frequency above which F_{opt} begins to deteriorate, is approximately equal to $f_\alpha/\beta^{1/2}$. This is the geometric mean of f_α and the frequency f_β at which β begins to fall, since $f_\beta = f_\alpha/\beta$. For operation at high frequencies it is therefore important that f_α should be as high as possible. Now f_α is determined largely by the total effective base to emitter capacity. To make this capacity small the junction area of the transistor should be small also, and this has several side effects. One of the factors that determines the value of β is the surface recombination effect. For β to be high it is important that the minority carriers should travel across the base from emitter to collector without combining with the majority carriers in the base region. Any recombination constitutes a base current that reduces the ratio of I_c to I_b and therefore of β. The probability of recombination is very much greater at the edge of the base region owing to surface imperfections, and to have a small junction area tends to increase the relative contribution of the surface recombination to the total base current. It is therefore rather more difficult to obtain a high value of β when the junction area is small, and the decrease of β with I_c becomes rather more rapid. In consequence I_c tends to optimize at higher values of I_c for high-frequency transistors than for low-frequency transistors. Values of 1 to 3 mA are typical. Another point to remember is that f_α itself is current-dependent to some extent. In fact

$$f_\alpha \simeq \frac{1}{2\pi r_e c_{b'e}} \tag{9.4}$$

where $c_{b'e}$ is the effective base emitter capacitance and is given by the expression

$$c_{b'e} = \frac{a_0 \tau_c}{r_e} + c_{te} \qquad (9.5)$$

Here $a_0 \tau_c / r_e$ represents the capacity associated with the minority carriers stored in the base region, τ_c the time needed for a minority carrier to traverse the base, and c_{te} the capacity associated with the separation of majority carriers on either side of the base emitter junction by the depletion layer. Since $r_e \sim I_c^{-1}$, it is clear that when I_c is large c_{te} can be neglected and $f_a \simeq 1/2\pi\tau_c$, i.e. f_a is independent of current. As I_c is decreased in order to improve the noise factor, r_e increases and so $a_0 \tau_c / r_e$ decreases, and finally f_a is given by $1/2\pi r_e c_{te}$. The value of c_{te} is independent of I_c and so f_a becomes proportional to I_c and begins to drop. This is another reason why I_c cannot be reduced very far at high frequencies.

As a rule, transistors that are designed for low-noise high-frequency operation will not work well at very low frequencies, and vice-versa. At very low frequencies, flicker noise becomes important and a large part of this noise originates from surface effects such as the surface recombination that has been mentioned. It is therefore advantageous from this point of view to have the largest area of junction possible. Then the surface recombination current will be low compared with the main part of the current flowing across the junction. To have a large junction area means a low value of f_a and so good low-frequency transistors are no use at high frequencies. On the other hand, good high-frequency transistors have a very small junction area and therefore a large amount of $1/f$ noise. The $1/f$ noise is, of course, of no consequence when the operating frequency is high, but when it is low the noise factor becomes very poor.

The problem of overcoming surface noise is now being achieved, to some extent, with silicon planar epitaxial transistors and for these the lower frequency noise corner, the frequency below which $1/f$ noise degrades the noise figure, can be quite low ($\simeq 100$ kc/s) for even very high frequency transistors.

Summarizing, we may say that, in general, one should pick the transistor that is suited to the frequency range in question, that the collector current, the resistive and the reactive matching should be optimized, and that for low frequencies I_c will optimize at lower values than it will for high frequencies.

100 KC/S PHASE-SENSITIVE DETECTOR

The choice of a phase-sensitive detector for the double-modulation spectrometer is difficult because of the vast number of circuits that are available. Broadly speaking, detectors fall into two classes, those with the switching type of circuit and those with what we shall term the 'additive' circuit. A simple example of the switching type of detector is shown in Fig. 9.5(a). This is a transistor switch with the reference signal fed to the base and the a.c. input signal to the collector. As the base circuit is biased on and off, a consideration of the Ebers and Moll equations[2] reveals that the collector-to-emitter junction is switched from being a near open circuit to being a near short circuit and back again. In one condition the input signal is therefore shorted and in the other it appears at the output terminal. Thus the switching action described in §2.3 is realized.

The additive-type phase-sensitive detector is shown in Fig. 9.5(b), and is nothing more than an ordinary envelope detector with two inputs, the reference and the signal. In one sense the selection of phase takes place prior to the detector and we can see this from Fig. 9.5(c). Here the phasor AB represents the reference signal and BC the input signal. Then $|AC|$, to which the detector responds, increases by $|BC| \cos \theta$, provided that $|AB| \gg |BC|$. The detection is therefore phase-sensitive.

About the most important characteristic of a phase-sensitive detector is its dynamic range. At the output of any such detector there is always a degree of drift and some $1/f$ noise, and it is important that the signal level should be high enough to make the drift and $1/f$ noise negligible in comparison. However, the signal can never be larger than the value at which the detector overloads, so the best ratio of signal to drift plus $1/f$ noise is the ratio of the overload level to the noise plus drift. This ratio is the dynamic range.

It will be immediately clear that the dynamic range of the simple additive detector is rather poor. The detector gives a steady output that is directly proportional to the reference signal, and any drift in the amplitude of that signal immediately appears at the output. The overload level is the level of signal that approaches the amplitude of the reference. The drift can be much reduced by using a balanced pair of detectors, with the phasing of the reference or the signal reversed between the two, but this is only a partial solution and is, anyway, a measure that is possible with any type of detector, good or bad.

It is one of the great advantages of the transistor switch that the reference signal results in little or no d.c. signal at the output of the detector. A close examination of the unfiltered output will reveal, in the absence of any input, a small square-wave signal with an amplitude in the region of a millivolt. The Ebers and Moll equations predict this offset voltage and indicate a value of $KT/q_E\beta_R$ for it. Now KT/q_E is 25 mV if T is room temperature; β_R represents the current gain of the transistor when it is operated with the collector and emitter reversed. Thus, because the reverse current gain of the transistor is seldom as high as the forward current gain, it is more usual to operate the transistor in the inverted mode, as shown in Fig. 9.6. Then it is the normal β that applies in the above expression and the offset is accordingly reduced. The overload level of the transistor switch is the value of the input signal that exceeds the reference voltage. Thus the dynamic range of the detector is better than the ratio of the maximum allowable base voltage to the value of $25\,\text{mV}/\beta$, a figure in the region of 60 to 70 dB. There are a few precautions to observe with this circuit, but provided that they are met this is probably equal to the best performance that can be obtained with any circuit so far developed. The first point to note is that the 'short circuit' resistance is given approximately by $kT/q_E I_B \cdot (\beta_R + \beta_F)/\beta_R\beta_F$, where β_F is the forward current gain. Now in many modern transistors the reverse current gain is very low indeed, and this can make the 'short circuit' resistance so high that the switch ceases to work. On the whole the older type of germanium junction transistor tends to be more symmetrical, giving a β_R that is only one tenth as low as the β_F, and works satisfactorily. A limitation of the transistor switch is that the normal effect of charge storage that determines the upper frequency limit of operation of any transistor generally degrades the performance at frequencies above 1 Mc/s. It is, unfortunately, not possible to use high-frequency transistors at these higher frequencies, because the techniques for raising the upper frequency cut-off of the transistor are also those that make the reverse current gain of the transistor much lower than the forward gain. One can do a good deal towards reducing the effects of charge storage by reducing the value of R_s, the collector resistor. However, R_s cannot be reduced beyond the point where it ceases to be large compared with the short circuit resistance of the transistor. It is generally this compromise that determines R_s. The value of R_B is quite simply the ratio of the maximum allowable base voltage to the maximum allowable base current.

A final point that is often missed is that the reference waveform must be a square wave and not a sine wave. If the reference is sinusoidal, the detector saturates for that part of the switching cycle where the switching waveform is passing through zero, and a gradually increasing non-linearity results. If a square wave is used, the linearity remains excellent up to the point where the signal amplitude is about to exceed the reference. Beyond this point, of course, overload occurs in the normal way.

The practical circuit adopted is shown in Fig. 9.6. Here a complementary pair of standard junction germanium transistors are used in a similar manner to that suggested above. There are two reasons for using the pair. One is that the drift, such as it is, from each transistor is compared in the difference amplifier following the detector and a degree of cancellation results. The other point is that, since the transistors are switched alternately, the impedance presented to the two blocking capacitors feeding the reference and input signals remains largely constant, and no charge is therefore built up on either capacitor. The 150 pF capacitors placed across the base resistors are included to compensate for the charge storage effects that otherwise tend to delay the switching of the transistor, and to give the same sort of distortion as results from a switching waveform that is not square. This measure is found to effect a considerable improvement in the dynamic range for an operating frequency of 100 kc/s.

The drift from the circuit is in the region of 100 μV per °C and it is therefore necessary to use a differential d.c. amplifier that

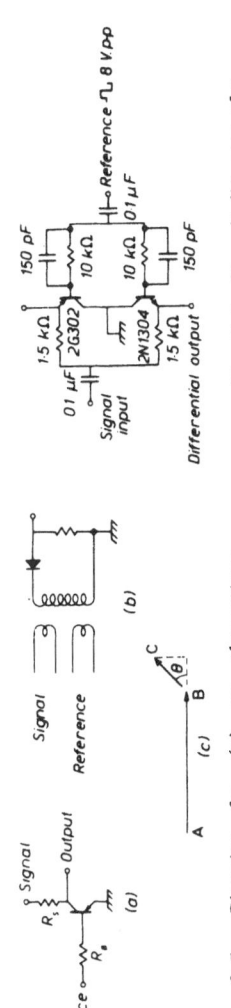

Fig. 9.5. Circuits for (a) an elementary switching phase-sensitive detector and (b) an elementary additive phase-sensitive detector. (c) Phasor diagram illustrating the operation of an additive phase-sensitive detector.

Fig. 9.6. Circuit diagram of a phase-sensitive detector.

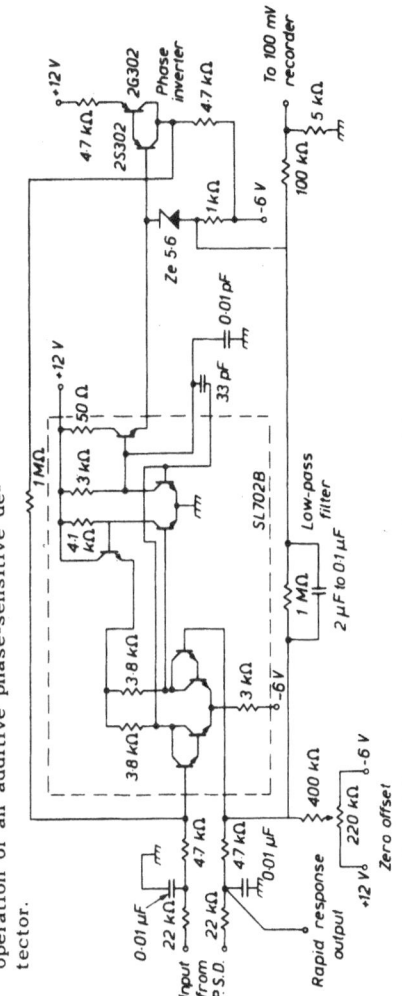

Fig. 9.7. Circuit diagram of a d.c. amplifier and a low-pass filter.

has comparable drift characteristics. Again a suitable integrated circuit operational amplifier is available. The circuit is the Plessey type SL702B shown in Fig. 9.7. This amplifier has a balanced input and a single output, an open loop gain of 70 dB and an input drift voltage of 30 μV per °C. Because of the single output, and in order to make the impedance presented to each half of the phase-sensitive detector equal, the operational amplifier is followed by a phase inverter. Balanced transadmittance feedback is then applied via the two networks comprising the 1 MΩ feedback resistors and the 22 kΩ + 4·7 kΩ input resistors to bring the gain down from the open loop value of 70 dB to about 32 dB. The resulting increase in gain stability and linearity is thereby made to be about 38 dB. The output from the amplifier is not symmetrical about earth and indeed is never negative. It is therefore necessary to incorporate the Zener diode to provide an output that does swing symmetically above and below zero. The output from the Zener diode is then attenuated to suit the recorder actually used, which has a full-scale deflection of 100 mV. The attenuation is adjusted so that the drift from the system is barely discernible on the recorder. The 26 dB loss of the output network, together with the 32 dB gain from the amplifier, caused the 100 μV/°C drift from the detector to be translated to 200 μV/°C on the recorder, that is 0·2 per cent of full-scale deflection. With these values the output from the amplifier has to swing over 2 V to give full-scale deflection of the recorder and this figure is well below the value at which the amplifier overloads.

The operational amplifier provides a convenient means of introducing the variable low-pass filter, which otherwise presents something of a problem. In the ordinary way the input impedance of a transistor amplifier is rather low and even the Darlington configuration of the input stages of the operational amplifier only raises the value to 30 kΩ. As a general rule it is desirable to have a filter with a variable time response from 0·1 sec to a few seconds for chart recording, and to have a response comparable with the bandwidth of the tuned amplifier for rapid response applications such as when the field of the magnet is being swept at 50 c/s. In the ordinary way capacities in the region of several hundreds of microfarads would be required to achieve the longer time constants with the 30 kΩ input impedance. Electrolytic capacitors would therefore have to be used and the associated leakage when these are placed at the input of the amplifier results in a drift that vastly degrades the overall drift characteristics of the system. It is therefore fortunate that a 2 μF capacitor (about the largest, conveniently sized, non-electrolytic capacitor) placed across one of the 1 MΩ feedback resistors gives a time constant of 2 sec, which is about the longest value ever really needed. Suitable reduction of the capacity can then give the desired range of response times. Because of the tendency of noise to overload the amplifier when the capacitor is removed, it is better to take the output for fast response applications directly from one of the 0·01 μF smoothing capacitors at the output of the detector. The effective cut-off frequency at this point is determined by the 0·01 μF capacitor and, because the input impedance to the amplifier with feedback is extremely low, by the 4·7 kΩ resistor. The cut-off frequency is therefore 3 kc/s, which is comparable with one half of the bandwidth of the tuned amplifier. A zero offset control is incorporated at the input to the amplifier.

It is generally necessary to reduce the cut-off frequency of the operational amplifier by inserting a capacitor between the two terminals brought out of the microcircuit for this purpose. For normal values of negative feedback fraction, a value of 33 pF is recommended to avoid oscillation when feedback is applied. With the present arrangement the input and output are virtually shorted together, and the 33 pF capacitor has to be supplemented by the 0·01 μF capacitor shown.

It is of value to note that the dynamic range of the phase-sensitive detector gives a direct indication of the degree to which the output of the detector can be usefully filtered to reduce the output noise level. If the input noise level is just less than the overload value, then, because the output noise voltage varies with the square root of the bandwidth of the low-pass filter at the output, the output noise will fall to the level of the output drift and $1/f$ noise when the bandwidth of the output filter is equal to the half-bandwidth of the preceding tuned amplifier divided by the square of the dynamic range. Thus, for a 10 kc/s input bandwidth and a 60 dB dynamic range, the output bandwidth can usefully only be reduced to 0·005 c/s, i.e. to give a time constant of about 30 sec.

§9.5

100 KC/S OSCILLATOR

We shall now turn to the section of the spectrometer that provides the 100 kc/s field modulation and the reference signal for the phase-sensitive detector. The first circuit to consider is the 100 kc/s oscillator. The requirements for this are a moderate degree of frequency stability, a constant amplitude output, and a fairly sinusoidal waveform. The frequency should be sufficiently stable to prevent the phase of the signal changing in time as it passes through the tuned 100 kc/s amplifier. The amplitude should be constant since this directly affects the sensitivity of the spectrometer. Finally, the waveform should be sinusoidal because the operation of the phase shifter is based on this assumption.

The practical circuit is shown in Fig. 9.8. A quartz crystal is used to determine the frequency of oscillation because this gives an excellent frequency stability and the circuit is not required to be tunable. The feedback loop comprises a gain-stabilized amplifier, a diode squarer and the quartz crystal. The diode squarer

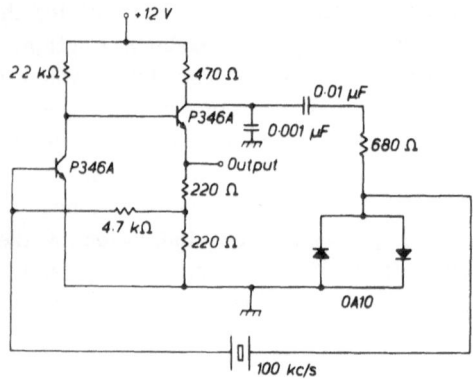

Fig. 9.8. Circuit diagram of a 100 kc/s oscillator.

determines the amplitude of oscillation by providing a constant amplitude output that is fed to the crystal. The crystal filters this signal to provide a sinusoidal output of a similarly constant amplitude. Then the gain-stabilized amplifier raises the level of the output from the crystal. The final output is therefore of constant amplitude, almost perfect waveform and of ample frequency stability.

The value of output voltage is 1·2 V peak-to-peak and this changes by only 6 per cent for a 50 per cent change in supply voltage.

§9.6

100 KC/S PHASE SHIFTER

The phase shifter is required to have a phase shift from zero to at least 180°. As usual the requirement is that the gain of the phase shifter should not vary with time, but it is also important, from the point of view of ease of adjustment, that alteration of the phase-shift control should not alter the amplitude of the output signal.

The practical arrangement is shown in Fig. 9.9 and works as shown in Fig. 9.10. The principle is to provide a phase-split voltage A and B across which a simple series *r.c* network is placed. The signal between the centre point C of the network and the earth then varies in phase as the ratio of *c* to *r* is varied, without altering in amplitude. Thus the result can easily be obtained by complex algebra but it is also simply illustrated by the vector diagram of Fig. 9.10(*b*). Here EA and EB represent the phase-split driving voltage, whilst AC and CB represent the voltages across *r* and *c* respectively. Since the current through *r* and *c* is common, AC and CB must be in quadrature, so when either *c* or *r* is varied C traces a half circle. The output voltage is CE, which therefore varies in phase but not in amplitude.

It will be clear from the above picture that the absolute maximum phase shift that can be obtained by altering the ratio of *c* to *r* is 180°. Thus to obtain a comfortable 180° it is advisable to use two phase-shifting sections. In Fig. 9.9 the first transistor splits the phase of the 1·2 V peak-to-peak output signal from the crystal oscillator. The 470 pF capacitor has a reactance of 3 kΩ at 100 kc/s, whilst the 10 kΩ potentiometer allows the effective output impedance of the upper half of the phase splitter to vary from 1 kΩ to 11 kΩ, a figure of one third to about three times the reactance of the capacitor. The phase change

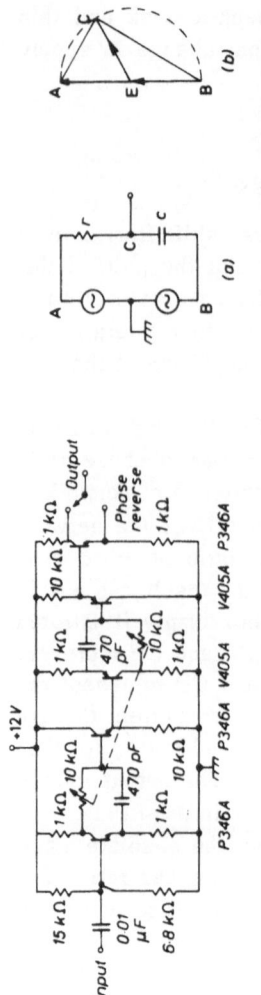

Fig. 9.9. Circuit diagram of a 100 kc/s phase shifter.

Fig. 9.10. (a) Functional diagram of a phase shifter. (b) Associated phasor diagram.

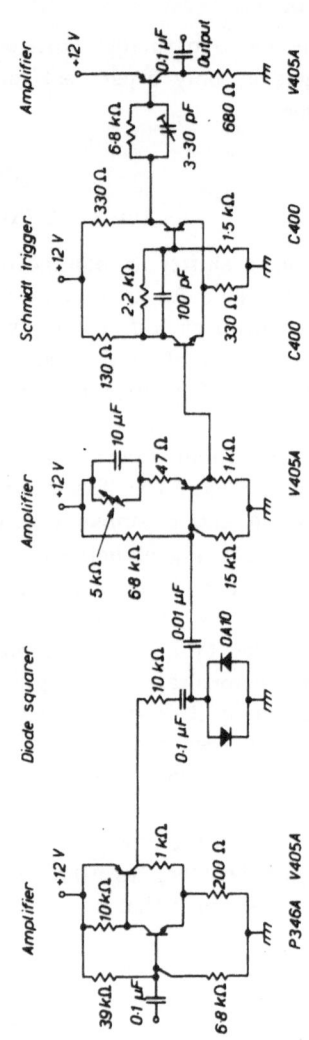

Fig. 9.11. Reference squaring and amplifying circuits.

thereby obtained is well over 90°, so with two similar stages the required phase shift is easily obtained. The above theory assumes no loading of the centre point of *c* and *r*, and it is therefore necessary to precede the next phase splitter, the first V405A transistor, by an emitter follower to raise the loading impedance to a value well in excess of 3 kΩ. It is similarly necessary to precede the final phase splitter by an emitter follower. The final phase splitter and switch is provided to allow complete phase reversal, a facility that is sometimes useful. The emitter followers are made complementary to the phase splitters that they precede in order to cancel the base-to-emitter voltage drop that otherwise builds up to an inconveniently large figure over five successive stages. A phase shift variation of 140° per stage is obtained in practice, with only a 6 per cent overall change in amplitude. A 10 per cent change in the d.c. supply voltage produces only a 2 per cent change in the output amplitude. The output voltage is 0·8 V peak-to-peak for an input voltage of 1·2 V. The circuits are therefore operating considerably below the overload point.

§9.7

100 KC/S SQUARER

The remaining section of circuitry between the phase shifter and the phase-sensitive detector is the squarer and amplifier that converts the fairly low level sinusoidal output from the phase shifter to a suitably sharp square wave of the amplitude required to drive the phase-sensitive detector. The squarer and amplifier are shown in Fig. 9.11. The first two transistors comprise a conventional voltage feedback pair that raises the amplitude of the reference sine wave to 4 V peak-to-peak. The diode squarer then converts the sine wave to a rough square wave which is amplified by a simple single-transistor amplifier to an amplitude of 4 V peak-to-peak again. There then follows a Schmidt trigger circuit that sharpens the rough input into a nearly ideal square wave of 3 V peak-to-peak amplitude. The 5 kΩ resistor controls the operating point of the transistor amplifier and so determines the mark-to-space ratio of the output from the trigger circuit. The final amplifier raises the reference amplitude to 8 V peak-to-peak, which is the value suitable for the phase-sensitive detector. Charge storage compensation is afforded by the trimming capacitor in the base circuit of the final transistor. This capacitor is adjusted to give the best square wave at the output.

§9.8

100 KC/S POWER AMPLIFIER

The initial arrangement for providing the 100 kc/s power to drive the field modulation coils is shown in Fig. 9.12. This system

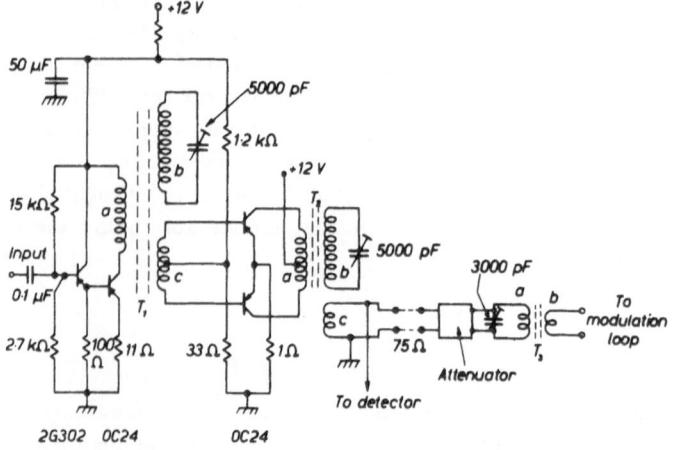

Fig. 9.12. Basic power amplifier for providing the cavity with a field modulation of 100 kc/s.

T_1: $a = 5$ turns T_2: $a = 4$ turns centre tapped
  $b = 39$ turns $b = 39$ turns
  $c = 8$ turns centre tapped $c = 4$ turns
T_3: $a = 80$ turns
  $b = 10$ turns
T_1, T_2 and T_3 are on Vinkor LA2106 formers.

comprises a low-power emitter-follower to provide some current gain, then a medium-power tuned phase splitter, and finally a push-pull class AB power amplifier. A single-turn modulating loop provides the field modulation in the cavity. In order to obtain the most efficient transmission, and also to provide a proper match for the output attenuator, a tuned transformer is arranged to present an impedance of 75 Ω to the 75 Ω line that feeds the modulating signal to the cavity. It was found, however,

that such an arrangement only gives an output amplitude stability of 6 per cent, and this is decidedly the limiting factor in
determining the sensitivity stability of the spectrometer. In order
to improve the stability the stabilizing system shown in outline

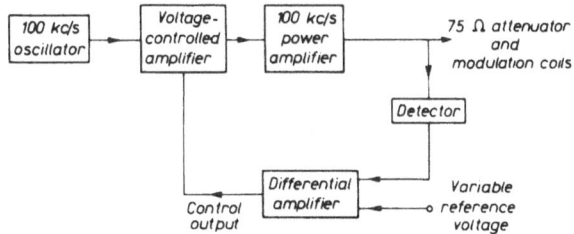

Fig. 9.13. System for stabilizing and controlling the amplitude
of the field modulation.

in Fig. 9.13 was therefore adopted. Here the input signal to the
75 Ω attenuator is detected and the d.c. output from the detector
is compared with a d.c. reference voltage in a high-gain differential amplifier. The output from the amplifier is then used to
control the gain of a further stage that is connected between the
100 kc/s oscillator and the power amplifier. In this way the output of the amplifier is stabilized so that the detected d.c. output
is equal to the d.c. reference voltage. A further advantage of the
stabilizing system is that the power needed to drive the modulation coils to the maximum extent required is about 4 W. This
value is a little high for the average attenuator and it is more
convenient to control the output by varying the reference voltage.
Control of the modulation in this way is adequate over about a
decade of field. For lower values the detector ceases to operate
properly but by then the power has fallen by two decades and the
attenuator can be used without difficulty.

§9.9

100 KC/S GAIN CONTROLLED AMPLIFIER

The circuit diagram of the gain-controlled amplifier inserted between the crystal oscillator and the power amplifier is shown in
Fig. 9.14. The method of controlling the gain is to provide a low-
impedance source and to use a variable base bias to control the
input impedance of the transistor. Input impedance is inversely

proportional to base current, so, assuming a constant current gain, the a.c. output current from the transistor is proportional to the d.c. base current. The input signal from the 100 kc/s crystal oscillator is attenuated by 34 dB, using the input potentiometer shown. This partly lowers the source impedance in the required manner, and partly reduces the amplitude of the input signal enough to avoid distortion. Operation of a transistor amplifier

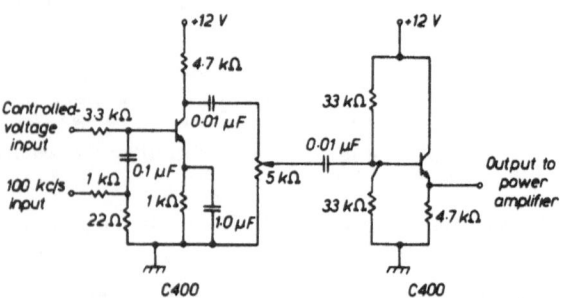

Fig. 9.14. Circuit of a gain-controlled amplifier.

with a constant voltage source introduces distortion far more readily than does operation with a constant current source. Checks reveal that the distortion is negligible and that electronic variation of the gain introduces a negligible change in phase shift. The emitter follower is added to provide isolation, and the potentiometer to allow the mean gain of the system to be set to unity. The value of $3 \cdot 3$ kΩ for the voltage-control source impedance has to be chosen rather carefully. In the ordinary way, increasing the base bias causes the input impedance to drop and the gain to increase. If, however, the transistor bottoms the gain begins to fall very sharply. The sense of the feedback is then reversed and the control loop locks into an unstable state. The value of the resistor is chosen so that the differential amplifier that provides the control voltage overloads before the control transistor bottoms.

§9.10

100 KC/S FIELD MODULATION DETECTOR AND AMPLIFIER

The remainder of the system for stabilizing and controlling the amplitude of the field modulation is shown in Fig. 9.15. The detector is of conventional design and gives an output determined by the amplitude of the voltage across the input to the 75 Ω attenuator. The d.c. output is added to the variable reference voltage, which is arranged to be of opposite sign, and fed to one input of the differential amplifier. Once again the Plessey type SL701B integrated-circuit operational amplifier is suitable for use as the differential amplifier. The output from the amplifier is then connected to the gain-controlled 100 kc/s amplifier, as already discussed. The 25 kΩ gain control is not essential and is

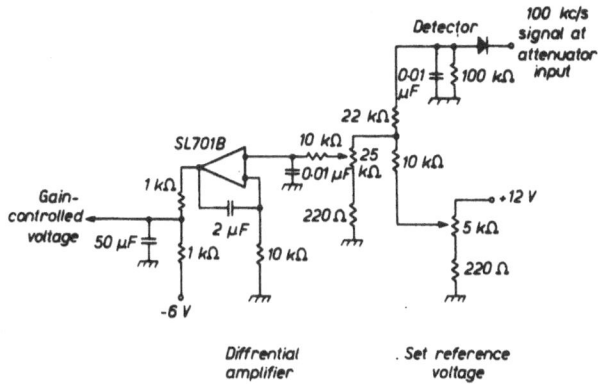

Fig. 9.15. Circuits for detecting a 100 kc/s field modulation, comparing it with a reference voltage and for amplifying an error signal for gain control.

normally set to its maximum value. It is useful, however, for making open-loop measurements on the system, when it can be set to its lowest value, the loop being broken at the output of the detector and the then much attenuated reference voltage being used as an input. The main time constant of the control loop is afforded by the 2 μF feedback capacitor and the 10 kΩ input resistor. This time constant is increased by the 70 dB open-loop gain of the amplifier to about 1 minute. The effective gain of the amplifier drops from the value of 70 dB as the frequency increases, and becomes constant at a value of unity at a frequency of 7 c/s determined by the 2 μF capacitor and the 10 kΩ resistor. The 50 μF capacitor at the output of the amplifier is chosen, together with the associated resistors, to provide a further cut-off at this

frequency so that the drop continues. This capacitor and the
0·01 μF capacitor at the input of the amplifier are necessary to
prevent an overall 100 kc/s feedback around the loop from occurring. The overall zero-frequency loop gain of the system is over
100 dB, which is more than adequate.

§9.11

10 KC/S A.F.C. CIRCUITS

It is clear from Fig. 9.1 that the individual circuits required for
the 10 kc/s A.F.C. system are very much the same as those for
the 100 kc/s system, namely, an oscillator, a tuned amplifier,
a phase-sensitive detector, a phase shifter and a d.c. amplifier.
The principle difference in each circuit, apart from the frequency,
is that the gain stability requirements are not so exacting. In
consequence a smaller degree of feedback is necessary and one
can economize by using fewer transistors.

§9.12

10 KC/S OSCILLATOR

The 10 kc/s oscillator shown in Fig. 9.16 is of conventional de-
design. The output is divided by two separate output potentiometers and must be of suitable amplitude to supply the phase
shifters and the modulation to the klystron reflector. The amplitude
of the latter is arranged to be 0·1 V peak-to-peak which is enough
to modulate the klystron by the equivalent of 100 kc/s. This figure is not enough to degrade significantly the resolution of the
spectrometer since the frequency of field modulation is also
100 kc/s.

§9.13

10 KC/S PHASE SHIFTER AND SQUARER

The circuit diagrams of the 10 kc/s phase shifter and squarer are
shown in Fig. 9.17. The phase shifters are designed on similar
lines to the 100 kc/s circuits, but the necessary isolation between
the stages is afforded by using 100 per cent voltage feedback pair
circuits. This arrangement is slightly simpler than the 100 kc/s

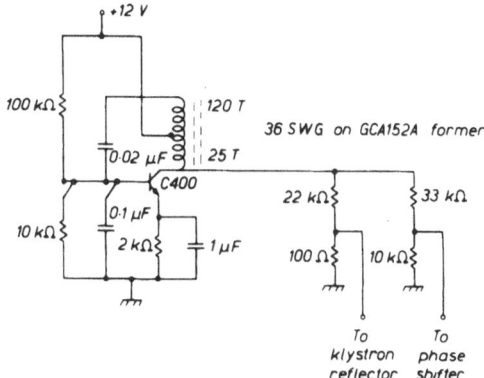

Fig. 9.16. Circuit diagram of a 10 kc/s oscillator

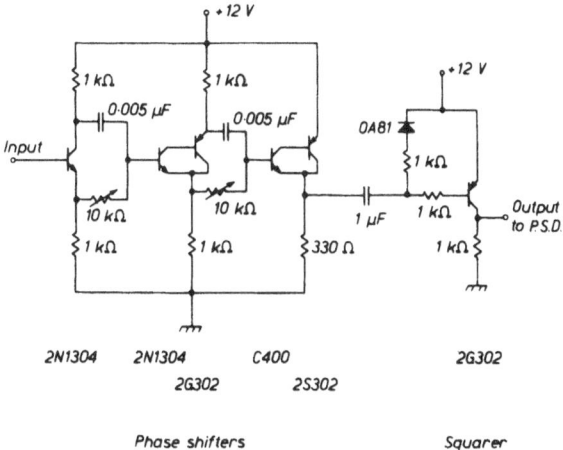

Fig. 9.17. Circuit diagram of a 10 kc/s phase shifter and squarer.

circuit and might well also have been used in this instance. Slight
nonlinearity in the operation of the 10 kc/s phase-sensitive
detector is not such a serious matter as it is in the 100 kc/s system, so a simple overloaded transistor is used as a squarer. The
OA81 diode and a 1 kΩ resistor maintain a 1 kΩ load on the 1μF
blocking capacitor when the transistor is not drawing base current, and so prevents a d.c. charge from building up on the
capacitor.

§9.14

10 KC/S PHASE-SENSITIVE DETECTOR

The circuit diagram of the 10 kc/s phase-sensitive detector is
shown in Fig. 9.18. The design is exactly the same as the 100
kc/s circuit except that the resistors have slightly lower values.
The values in Fig. 9.18 were arrived at after more careful consideration and are generally more suited to the transistors in
question. The 100 kc/s circuit was judged to be adequate however and was not altered.

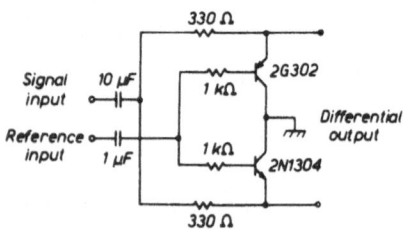

Fig. 9.18. Circuit diagram of a phase-sensitive detector.

§9.15

10 KC/S TUNED AMPLIFIER

The circuit diagram of the 10 kc/s tuned amplifier is shown in
Fig. 9.19. All the stages are of conventional design, and negative
feedback is applied to make the gain of each 20 dB. In this way
the gain remains substantially constant with drift in the β of the
transistor until the value of β falls below 10, an unlikely event-
uality. The bias is arranged to make the collector current of all
stages but the last equal to 1 mA. The load to each stage is determined largely by the 3·3 kΩ bias resistor of the next stage,
because only when β falls below 10 does the effective input impedance fall to below 3 kΩ. The input impedance of the trans-
istor is equal to β 25 mV/I_e or > 250 Ω for (β > 10) plus βR_E,
where R_E is the resistor in the emitter lead, i.e. 3 kΩ. The
ratio of 10 : 1 for the load-to-emitter resistor then determines the
20 dB gain figure. The first two stages are tuned and each
tuned circuit is designed to have a Q-factor of 2. The combined
frequency response of these, together with that of the Wien network at the input, which has a Q of 1/3, gives the widest pos-

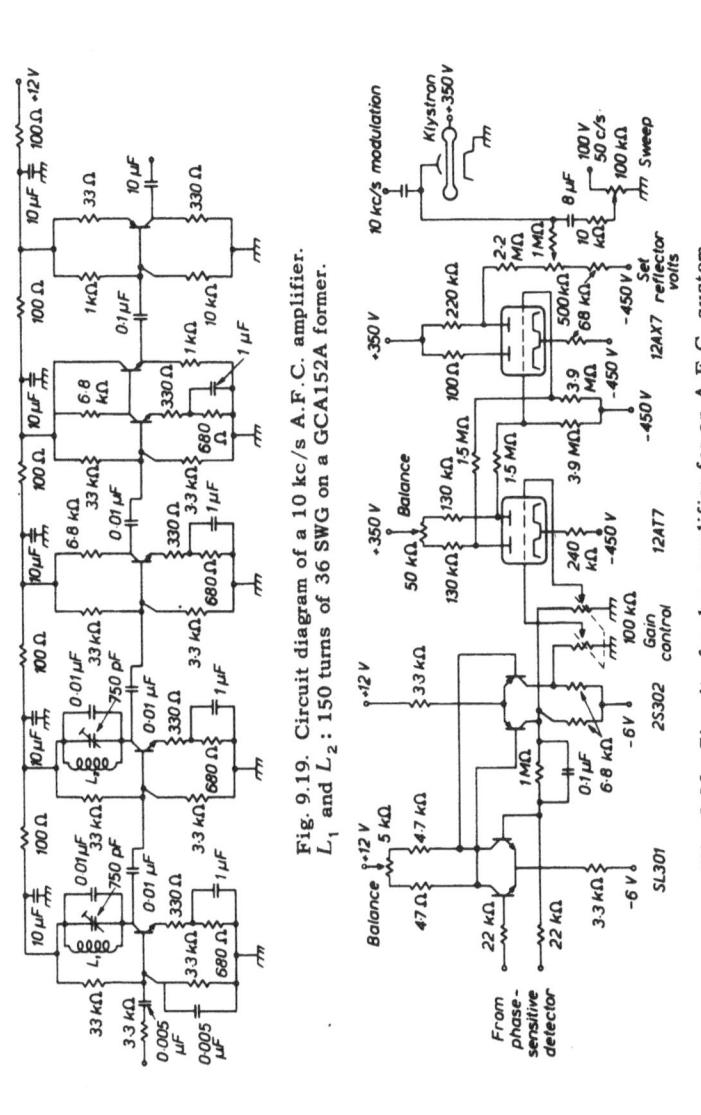

Fig. 9.19. Circuit diagram of a 10 kc/s A.F.C. amplifier.
L_1 and L_2: 150 turns of 36 SWG on a GCA152A former.

Fig. 9.20. Circuit of a d.c. amplifier for an A.F.C. system.

sible bandwidth whilst giving adequate rejection at all harmonics of the operating frequency. A wide bandwidth is necessary because the system is in a feedback loop and, apart from the main time lag in the system, should have the most rapid response possible. The final stage has to draw rather more current than the others in order to have a sufficiently large current output to drive the phase-sensitive detector without overloading itself. The correspondingly lower input impedance of the stage demands an emitter follower between it and the previous stage. The final number of stages was determined empirically to be nearly enough to overload the phase-sensitive detector with noise if need be.

§ 9.16

A.F.C. D.C. AMPLIFIER AND KLYSTRON CONTROL CIRCUITS

The circuits for amplifying the error signal and feeding it to the klystron reflector are shown in Fig. 9.20. The problem of coupling the output of a d.c. amplifier to the reflector of a klystron is always a somewhat awkward one because the potential of the reflector is so much below that of the resonator, which normally has to be earthed since it is connected to the main body of the waveguide. The method that I favour is one which requires the cathode of the klystron, rather than the resonator, to be earthed, and a d.c. break to be placed between the klystron output and the remainder of the waveguide. The obvious drawback of this arrangement is that it leaves the body of the klystron 'alive'. This is not however an insuperable problem. Sufficient caution and insulation of the tuning knob can prevent the system from being too dangerous. The advantage of the arrangement is that both the d.c. filament supply for the klystron and the d.c. amplifier do not have to float at a potential other than earth, and the additional problem of arranging for a drop in the direct current to occur somewhere in the 10 kc/s section of the circuitry is neatly avoided. The first stage of the d.c. amplifier is a matched pair of transistors on one integrated circuit chip. The transistor pair is the Plessey SL30IA, which has an input drift coefficient of $7\mu V/^{\circ}C$ and is therefore rather better than the phase-sensitive detector that precedes it. The number of stages that can usefully be added after the phase-sensitive detector is determined by the normal klystron drift. It is advantageous to increase the gain of the d.c. amplifier only up to the point where the drift originating from the phase-sensitive detector exceeds the original drift of the klystron. The remaining consideration is to adjust the main time lag in the system to be long enough to prevent hunting of the loop. The longest time constant that can be conveniently used is formed by the 8 μF capacitor and the 1 MΩ resistor in the reflector circuit. This circuit represents a cut-off frequency of about 0·01 c/s whilst the potential cut-off frequency of the remainder of the loop is in the region of 1 kc/s. A loop gain of 10^5 is therefore possible, being the ratio of these two frequencies. If the 8 sec time constant is simply placed at the output of the d.c. amplifier when this value of gain is used, which is in fact approximately what the loop affords, the later stages of the d.c. amplifier overload with noise. If, on the other hand, the long time constant is placed at the input of the d.c. amplifier, hum etc. picked up in the front end of the d.c. amplifier are transferred directly to the reflector. As a compromise therefore, the 0·1 μF capacitor is placed across the 1 MΩ feedback resistor that stabilizes the gain of the first two stages of the d.c. amplifier. This gives an effective cut-off of 1·4 c/s at the input of the d.c. amplifier, and the noise is thereby prevented from overloading the system. Then to prevent hunting of the loop, the 20 dB decade drop due to the 8 sec time constant is checked at approximately the same frequency (1·4 c/s) by inserting a 10 kΩ resistor in series with the 8 μF capacitor.

A 50 c/s sweep for setting up purposes can be applied to the reflector in the manner already discussed. The 10 kc/s modulation is fed to the reflector by a small blocking capacitor.

Valves have to be used in the final stages of the d.c. amplifier to provide an adequate output to correct for the probable changes in the klystron frequency. The arrangement used gives an output that is nearly enough to traverse a single mode of the klystron before the d.c. amplifier overloads.

The method of adjusting and operating the complete A.F.C. system is as follows. First, the two d.c. balance potentiometers are adjusted and then left alone in subsequent use. The first balance control is set so that for zero output from the phase-sensitive detector the differential output from the 2S302 pair is zero. With the twin 100 kΩ gain control set to zero, the balance control in the anode circuit of the 12AT 7 is then adjusted to bring the second half of the 12AX7 to a point half way between cut-off and bottoming.

In use, the gain control is first left at zero and the 50 c/s sweep is applied to the reflector via the 100 kΩ sweep potentiometer. The klystron mode pattern is displayed on the oscilloscope in the usual manner and the klystron cavity adjusted to bring the response curve of the spectrometer cavity exactly in the centre of the klystron mode. The sweep voltage is then gradually reduced and the reflector voltage set by the 500 kΩ 'set reflector' potentiometer to keep the klystron tuned exactly to the centre of the resonance curve of the spectrometer cavity. When the sweep is completely removed, the A.F.C. gain control is advanced to lock the klystron frequency to the centre of the resonance curve.

§ 9.17

FINAL PERFORMANCE

Final checks on the complete spectrometer reveal that the drift in sensitivity of the entire system is better than 1 per cent 10 minutes after switching on. A day-to-day check revealed a stability of 1 per cent also. The performance of the A.F.C. system is such that, provided that the klystron cavity is set so that the klystron operates at mode centre, no detectable asymmetry is observable in the spectrum traced from a standard carbon sample. With the automatic control of the modulation amplitude, no variation is discernible on a meter connected across the output of the sampling detector 10 minutes after switching on.

§ 9.18

EARTH LOOPS

It is probably worth while in closing to discuss the d.c. breaks that are frequently found to be necessary between certain sections of the waveguide in the spectrometer. Introduction of such a break between the microwave detector and the rest of the waveguide can, for example, prevent a strong pick-up of radiation from the 100 kc/s power amplifier by the 100 kc/s tuned amplifier. The usual method of making such a break is to place a thin sheet of mica between the two waveguide flanges required to be separated. The normal metal coupling rings are replaced by similar rings made of insulating material. Then the microwave signal can pass from one section of the waveguide to the next without there being any d.c. connection.

The mechanism for the pick-up that the d.c. break overcomes is illustrated in Fig. 9.21. The chassis and earth terminal of the 100 kc/s amplifier are connected to the main body of the wave-guide of the spectrometer by two paths, one via the outer lead of the coaxial cable connecting the amplifier to the detector, and the other, because both the waveguide and the amplifier are connect-

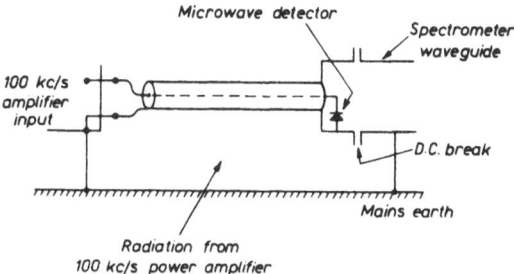

Fig. 9.21 Diagram illustrating the mechanism of an 'earth-loop' pick-up.

ed, to the mains-supply earth via the power supply to the ampli-fier and the power supplies to other parts of the electronics in contact with the waveguide. In consequence an 'earth-loop' is formed. If any generator is radiating a small amount of alternating magnetic field, this field induces an e. m. f. into the loop and, because it is shorted, heavy currents flow around it. The most powerful generator in the spectrometer under consideration is the 100 kc/s power amplifier and from it a small degree of radiation is inevitable. Thus a significant current at 100 kc/s flows round the earth-loop. Part of this loop is common with the loop formed by the inner and outer conductors of the coaxial cable, the input to the 100 kc/s amplifier and the output terminals of the micro-wave detector. In consequence, a 100 kc/s e.m.f. is induced in this second loop and part of the voltage appears across the input to the amplifier. If, however, the first loop is broken by separating the waveguide at the point suggested, the initial current cannot flow, so no signal is coupled into the second loop.

A measure sometimes taken in error is to break the outer connector of the coaxial cable. It will by now be clear that this measure simply leaves a loop comprising the inner conductor of the cable and the mains-earth return path, and the situation is probably made worse rather than better.

References

1. Nielsen, E.C., *Proc. I.R.E.*, 1957, **45**, 839.
2. Ebers, J.J., and Moll, J.L., *Proc. I.R.E.*, 1954, **42**, 1761.

BIBLIOGRAPHY

General

Chestnut, H., and Mayer, R.W., *Servomechanisms and Regulating System Design* (Chapman & Hall, Vol. 1, 1952; Vol. 2, 1955).

Fraser, W., *Telecommunications* (MacDonald, 1957).

Pound, R.V., *Microwave Mixers* (McGraw Hill, 1948).

Ramo, S., and Whinnery, J.R., *Fields and Waves in Modern Radio* (Wiley, 1953).

Electron Spin Resonance

Assenheim, H.M., *Introduction to Electron Spin Resonance* (Hilger & Watts, 1966).

Carrington, A., and McLachlan, A.D., *Introduction to Magnetic Resonance* (Harper and Row, 1967).

Ingram, D.J.E., *Free Radicals* (Butterworths, 1955).

Low, W., *Paramagnetic Resonance in Solids* (Academic Press, 1960).

Pake, G.E., *Paramagnetic Resonance* (Benjamin, 1962).

Poole, C.P., Jr, *Experimental Techniques in Electron Spin Resonance* (Wiley, 1967).

Siegman, A.E., *Microwave Solid-State Masers* (McGraw Hill, 1964).

Slichter, C.P., *Principles of Magnetic Resonance* (Harper and Row, 1963).

GLOSSARY

Vectors are used explicitly very rarely. Where vectors occur they are given in bold type, e.g. **H**. Generally an italic capital is used and then it represents the magnitude of the vector, e.g. H.

For scaler quantities, e.g. voltage, v represents the instantaneous value, $\hat{v}$ the peak value, $\tilde{v}$ the r.m.s. value, and V the phasor. Thus $\hat{v} = |V|$ and $\tilde{v} = (\overline{v^2})^{\frac{1}{2}}$.

A	Voltage gain of d.c. amplifier in an A.F.C. loop $[= k_A G_A(\omega)]$
A_n	Spectral intensity $W(f)$ of $1/f$ noise $= A_n/f$
A_s	Height of a voltage step
a	Real part of V_L; area
a_s	Area of E.S.R. sample
a_w	Area of cavity wall
B	Noise bandwidth; magnetic flux density
b	Imaginary part of V_L
C	Capacity in equivalent circuit of microwave cavity
c	Capacity per unit length of transmission line; speed of light
$c_{b'c}$	Total effective base-to-emitter capacity for a transistor
c_{te}	Part of $c_{b'e}$ associated with depletion layer
D_p	Pole diameter for magnet
E	Energy
E_g	E.M.F. of generator associated with equivalent circuit of klystron
E_S	E.M.F. of spin-generator associated with E.S.R. sample
E_R	Component of E_S in phase with applied microwave field
E_X	Component of E_S in quadrature with applied microwave field
$E_{1,2}$	Energy levels
e	The exponental function
F	Noise figure
$F_{1,2}$	Noise figures of cascaded amplifiers

F_{opt}	Optimum value of F
f	Frequency of a generator, most commonly the klystron
f_a	Frequency at which klystron is modulated for A.F.C. purposes
f_c	Cut-off frequency of low-pass filter
f_d	Frequency deviation when frequency is modulated
f_E	Frequency of E.S.R.
$f_{i.f.}$	Intermediate frequency of superheterodyne
f_K	Klystron frequency (usually f)
f_m	Frequency at which field of spectrometer magnet is modulated
f_n	Frequency of a noise component
f_o	Resonant frequency (usually of microwave cavity)
f_R	Cut-off frequency of rth stage in amplifier of gain A
f_α	α cut-off frequency for transistor
f'	Frequency component of a transient
G	Power gain
G_a	Power gain of low-noise pre-amplifier
$G_{1,2}$	Gains of cascaded amplifiers
$G(\omega)$	See $kG(\omega)$
$G_A(\omega)$	See A
$G_{ar}(\omega)$	See $k_{ar}G_{ar}(\omega)$
H	Magnitude of microwave field in cavity
H_{max}	Spatial maximum value of H
H_o	Field of spectrometer magnet
H_{sat}	Microwave field strength at which E.S.R. sample saturates
H_+	Magnitude of rotating microwave field
h	Planck's constant, $6 \cdot 625 \times 10^{-34}$ joule sec
$h(t)$	Chopping function
$h_s(t)$	Reversing function
$\hbar$	H/H_{max}
I_b	Base current of transistor
I_c	Collector current of transistor
I_e	Emitter current of transistor
I_o	Surface current associated with skin depth
i_n	Noise current
j	$\sqrt{-1}$

$K_{u,s}$	Constants defined by equations (4.21) and (4.22)
k	Boltzmann's constant ($1 \cdot 38 \times 10^{-23}$ joule/$^{\circ}$C)
$kG(\omega)$	Loop gain of A.F.C. system
$k_A G_A(\omega)$	Gain of d.c. amplifier in A.F.C. system ($= A$)
k_A	Zero frequency value of $k_A G_A(\omega)$
$k_{AR} G_{AR}(\omega)$	Gain of rth stage in d.c. amplifier in A.F.C. system
k_{AR}	Zero-frequency value of $k_{AR} G_{AR}(\omega)$
$k_a G_a(\omega)$	Gain of a.c. amplifier in A.F.C. system
k_a	Mid-band value of $k_a G_a(\omega)$
$k_{ar} G_{ar}(\omega)$	Gain of rth stage in a.c. amplifier in A.F.C. system
k_{ar}	Mid-band value of $k_{ar} G_{ar}(\omega)$
k_{DIS}	Transfer function for microwave discriminator (rate of change of output voltage with input frequency)
k_K	Transfer function of klystron in A.F.C. system (rate of change of frequency with reflector voltage)
k_m	Efficiency of microwave discriminator in super-hetrodyne A.F.C. system
L	Inductance in equivalent circuit of microwave cavity
L_s	Length of cylindrical E.S.R. sample
l	Inductance per unit length of transmission line
M	Magnitude of oscillating magnetization; mutual inductance
M_o	Equilibrium d.c. magnetization
M_z	D.C. magnetization
$M_{1,2}$	Mutual inductances coupling in and out of microwave cavity
M_+	Magnitude of rotating magnetization
N_o	Noise power output
$N_{1,2}$	Equilibrium populations of energy levels 1,2
NTR	Noise temperature ratio
n_l	Number of lines in an E.S.R. spectrum
P_a	Power available from cavity due to E.S.R. sample
P_B	Bucking power
P_{BA}	Bucking power for A.F.C. detector
P_{BE}	Bucking power for E.S.R. detector
P_C	Power dissipated in cavity by klystron ($= P_i$)
P_{CA}	Power dissipated in cavity by klystron for A.F.C. system

P_{CE}	Power dissipated in cavity by klystron for E.S.R. system
P_h	Harmonic power
P_i	See P_C
P_n	Noise Power
P_n'	Noise power at input to E.S.R. detector due to fluctuations in the frequency of the klystron
P_o	Output power
Q	Quality factor of resonator (usually the microwave cavity).
Q_E	External Q factor
$Q_{E1,2}$	External Q factors for input and output couplings to transmission cavity
Q_{eff}	Effective Q factor due to thin cavity wall
Q_L	Loaded Q factor
Q_U	Unloaded Q factor
q	Charge
q_E	Charge on an electron
R'	Resistance in equivalent circuit of microwave cavity
R_a	Radius of cylindrical E.S.R. sample
R_c	Radius of cavity circle diagram
R_m	Resistance coupled into equivalent circuit of cavity by maser material
R_o	Waveguide impedance
R_s	Surface resistivity of waveguide wall
$R_{1,2}$	Values of R for twin-cavity spectrometer
r	Radius coordinate; resistance per unit length of transmission line
r_e	Intrinsic emitter resistance for transistor
r_x	Base spreading resistance for transistor
S	Dispersion suppression factor
S_i	Improvement in voltage signal-to-noise ratio obtained by using double-modulation recording
S_p	Magnet pole gap
S_s	Slowing factor for slow wave structure
(S/N)	Signal-to-noise ratio
T	Absolute temperature
T_a	Equivalent amplifier noise temperature

T_c	Cavity temperature
T_D	Decay time for resonator
T_f	Decay time for low-pass filter
T_o	Room temperature
T_{opt}	Optimum thickness of flat E.S.R. sample
T_p	Period of regular pulse train
T_r	Recording time
T_s	Thickness of flat E.S.R. sample
T_1	Longitudinal spin relaxation time
T_2	Transverse spin relaxation time
t	Time
V_B	Bucking voltage
V_D	Voltage input to microwave detector
V_{in}	Input voltage
V_o	Output voltage
v	Instantaneous voltage
$\hat{v}$	Peak voltage
$\tilde{v}$	R.M.S. voltage
v	Volume
v_g	Group velocity
W	Base width of transistor
W_f	Spectral intensity of noise
X	Reactance of equivalent circuit of microwave cavity
x	Cartesian coordinate; distance
x_c	Length of waveguide cell
x_o	Size of E.S.R. sample of uniform proportions roughly filling the cavity
y	Cartesian coordinate
Z_L	Load impedance presented to transmission line
z	Cartesian coordinate; cylindrical coordinate
α	Emitter-to-collector current gain of transistor unless otherwise defined
α_c	Coupling factor for directional coupler
$\alpha_{D'}$	Factor determining the frequency response of an amplifier
α_i	Improvement in voltage signal-to-noise ratio obtained by converting to superheterodyne detection
α_o	Zero frequency value of α

(For V_B through $\tilde{v}$: See notes at beginning of Glossary)

α_P	Attenuation constant for transmission line
α_2	See α_c
β	Bohr magneton; cavity coupling factor; base-to-collector current gain for transistor; other miscellaneous uses
β_F	Forward current gain β
β_P	Phase constant for transmission line
β_R	Reverse current gain β
$\beta_{1,2}$	Input and output coupling factors for transmission cavity
γ_P	Propogation constant for transmission line $(=\alpha_P + j\beta_P)$
Δ	Generally used as a prefix to indicate width, e.g. Δf = frequency bandwidth
ΔE	Splitting between energy levels
Δf	Frequency bandwidth
$(\Delta f)_a$	Bandwidth of tuned amplifier
$(\Delta f)_{ar}$	Bandwidth of rth stage of tuned amplifier
$(\Delta f)_E$	Frequency width of E.S.R. line
$(\Delta f)_{eff}$	Effective bandwidth (reduced) of cavity in nearly balanced bridge
$(\Delta f)_f$	Bandwidth of final low-pass filter at output of spectrometer
$(\Delta f)_{i.f.}$	Bandwidth of i.f. amplifier in superheterodyne
$(\Delta f)_L$	Loaded cavity bandwidth
$(\Delta f)_U$	Unloaded cavity bandwidth
$(\Delta f)_{UA}$	Unloaded cavity bandwidth for A.F.C. system
$(\Delta f)_{UE}$	Unloaded cavity bandwidth for E.S.R. spectrometer
ΔH	Magnetic linewidth of E.S.R. resonance line
ΔT	Pulse width
δ	Frequently used as a prefix to indicate a change, an addition to, or departure from norm in a parameter; skin depth
δf	$f - f_o$
δf_{CORR}	Correction applied to frequency of klystron by A.F.C. system
δf_K	Change in klystron frequency
δf_{KS}	Remaining error in klystron frequency after A.F.C. has been applied

δf_{KU}	Error in klystron frequency prior to application of A.F.C.
δf_n	Noise or drift in the detecting system of an A.F.C. loop referred to an imaginary variation in the reference frequency
δf_R	Real variation in the reference frequency of an A.F.C. system
δR	Addition to cavity resistance on account of E.S.R. absorption
δX	Addition to cavity reactance on account of E.S.R. absorption
δr	Addition to resistance per unit length of a transmission line on account of E.S.R. absorption
$\delta |V_B|$	Imagined drift in bucking voltage accounting for drift in the detecting system of an A.F.C. loop
$\delta |V_D|$	Imagined drift in input to detector of A.F.C. system accounting for output drift
δv_D	Change in output voltage from A.F.C. discriminator
δv_R	Change in the reflector voltage of a klystron
$\delta \rho_x$	Fluctuation in ρ_x on account of frequency fluctuation
ϵ	Dielectric constant
ϵ_0	Value of ϵ *in vacuo*
ϵ'	Real part of ϵ/ϵ_0
ϵ''	Imaginary part of ϵ/ϵ_0
η	Cavity filling factor
θ	Cylindrical coordinate
θ_R	Angle between E_R and V_L
θ_X	Angle between E_X and V_L
$\boldsymbol{\mu}$	Magnetic moment
μ	Magnetic permeability
μ_0	Free space value of μ
ρ	Voltage reflection coefficient from cavity
ρ_B	$2V_B/E_g$
ρ_D	$2V_D/E_g$
ρ_M	Value of ρ when cavity is matched but not necessarily on-tune
σ	Conductivity
τ_c	Transit time of charge carrier across base of transistor

ϕ	Cylindrical coordinate
ϕ_A	Phase shift introduced by d.c. amplifier in A.F.C. system
ϕ_{AR}	Phase shift introduced by rth stage of d.c. amplifier in A.F.C. system
ϕ_a	Phase shift introduced by a.c. amplifier in A.F.C. system
ϕ_{ar}	Phase shift introduced by rth stage of a.c. amplifier in A.F.C. system
ϕ_p	Phase between signal and reference of phase sensitive detector
χ	Susceptibility of E.S.R. sample to linearly polarized microwave field
χ'	Real part of χ
χ''	Imaginary part of χ
χ_+	Susceptibility of E.S.R. sample to circularly polarized microwave field
χ_+'	Real part of χ_+
χ_+''	Imaginary part of χ_+
ω	$= 2\pi f$ whatever the subscript of f; applies also to Δf and δf